Bowdoin Boys in Labrador

Jonathan Prince Cilley, Jr

BOWDOIN BOYS IN LABRADOR

An Account of the Bowdoin College Scientific Expedition to
Labrador Led by Prof. Leslie A. Lee of the Biological Department

by

JONATHAN PRINCE CILLEY, JR.

PREFACE

This letter from the President of Bowdoin College is printed as an appropriate preface to the pages which follow.

I thank you for the advanced sheets of the "Bowdoin Boys in Labrador. " As Sallust says, "In primis arduum videtur res gestas scribere; quod facta dictis sunt exaequanda. "

In this case, the diction is equal to the deed: the clear and vivacious style of the writer is fully up to the level of the brilliant achievements he narrates.

The intrinsic interest of the story, and its connection with the State and the College ought to secure for it a wide reading.

Very truly yours,
WILLIAM DEW. HYDE.

ON BOARD THE "JULIA A. DECKER, "
Port Hawkesbury, Gut of Canso,
July 6th. 1891.

Here the staunch Julia lies at anchor waiting for a change in the wind and a break in the fog. To-day will be memorable in the annals of the "Micmac" Indians, for Prof. Lee has spent his enforced leisure in putting in anthropometric work among them, inducing braves, squaws and papooses of both sexes to mount the trunk that served as a measuring block and go through the ordeal of having their height, standing and sitting, stretch of arms, various diameters of head and peculiarities of the physiognomy taken down. While he with two assistants was thus employed, two of our photographic corps were busily engaged in preserving as many of their odd faces and costumes as possible, making pictures of their picturesque camp on the side of a hill sloping toward an arm of the Gut, with its round tent covered with birch and fir bark, dogs and children, and stacks of logs or wood—from which they make the strips for their chief products, baskets—cows, baggage and all the other accompaniments of a comparatively permanent camp. They go into the woods and make log huts for winter, but such miserable quarters as these prove to be on closer inspection, with stoves, dirt and chip floor, bedding and food in close proximity to the six or eight inhabitants of each hut, suffice them during warm weather. We found that they elect a chief, who holds the office for life. The present incumbent lives near by St. Peter's Island, and is about forty years old. They hold a grand festival in a few weeks somewhere on the shore of Brasd'Or Lake, at which nearly every Indian on the Island is expected, some two thousand in all, we are informed, and after experiencing our good-fellowship at their camp and on board they invited us one and all to come down, only cautioning us to bring along a present of whiskey for the chief.

The Gut, in this part at least, is beautiful sailing ground, with bold, wooded shores, varied by slight coves and valleys with little hamlets at the shore and fishermen's boats lying off the beach. The lower part we passed in a fog, so we are ignorant of its appearance as though the Julia had not carried us within a hundred miles of it, instead of having knowingly brought us past rock and shoal to this quiet cove, under the red rays of the light on Hawkesbury Point, and opposite Port Mulgrave, with which Hawkesbury is connected by a little two-sailed, double-ended ferry-boat built on a somewhat famous model.

1

It seems that a boat builder of this place, who, by the way, launched a pretty little yacht to-day, sent a fishing boat, whose model and rig was the product of many years' experience as a fisherman, to the London Fisheries' Exhibit of a few years past, and received first medal from among seven thousand five hundred competitors. The Prince of Wales was so pleased with the boat, which was exhibited under full sail with a wax fisherman at the helm, that he purchased it and has since used it. Later, when the United States fish commission schooner Grampus was here with the present assistant commissioner, Capt. Collins, in command, the plans were purchased by our government on the condition that no copies were to be made without Mr. Embree's consent. A little later yet, a commissioner from Holland and Sweden came over, bought the plans and built a perfect copy of the original, the seaworthy qualities of which has caused its type to entirely displace the old style of small fishing boats in those countries. The boat's abilities in heavy waters have been tested many times, and have never failed to equal her reputation.

But, meanwhile, the Julia lies quietly at anchor, as if it were mutely reproaching your correspondent with singing another's praises when she has brought us safely and easily thus far, in spite of gales, fog, and headwind, calm, and treacherous tide, and even now is eagerly waiting for the opportunity to carry us straight and swiftly to Battle Harbor in the straits of Belle Isle, where letters and papers from home await us, and then up through the ice fields to Cape Chudleigh.

[The Real Start] Our real start was made from Southwest Harbor, Mt. Desert, the Monday after leaving Rockland. Saturday night, after a short sail in the dark and a few tacks up the Thoroughfare to North Haven village, we anchored and rested from the confusion and worry of getting started and trying to forget nothing that would be needed in our two and one-half months' trip. Sunday morning was nearly spent before things were well enough stowed to allow us to get under weigh in safety, and then our bow was turned eastward and, as we thought, pointed for Cape Sable. Going by the hospital on Widow's Island and the new light on Goose Rock nearly opposite it, out into Isle au Haut bay, we found a fresh northeaster, which warned us not to go across the Bay of Fundy if we had no desire for an awful shaking up. In view of all the facts, such as green men, half-stowed supplies and threatening weather, we decided that we must not put our little vessel through her paces that night, and chose the more ignominious, but also more comfortable course of putting into a harbor. Consequently after plunging through the rips off Bass Head, and cutting inside the big bell buoy off its entrance, we ran into Southwest Harbor and came to anchor. In the evening many of the party thought it wise to improve the last opportunity for several months, as we then supposed, to attend church, and to one who knew the chapel-cutting proclivities of many of our party while at Bowdoin, it would have been amusing to see them solemnly tramp into church, rubber boots and all. It is a fact, however, that every member of our party, with a possible exception, went to church in this place yesterday largely for the same reason.

Our little Julia rewarded our action of the night previous by taking us out by Mt Desert Rock at a rattling pace Monday morning, bowing very sharply and very often to the spindle-like tower on the rock, as she met the Bay of Fundy chop, and at the same time administered a very effective emetic to all but five or six of the Bowdoin boys aboard. She is wise as well as bold and strong, and so after nightfall waited under easy canvas for light to reveal Seal Island to our watchful eyes. Shortly after daylight the low coast was made out, the dangerous rocks passed, and Cape Sable well on our quarter. But there it stayed. We made but little progress for two days, and employed the time in laying in a supply of cod, haddock and pollock, till our bait was exhausted. Then we shot at birds, seals and porpoises whenever they were in sight, and from the success, apparently, at many when they were not in sight; put the finishing touches on our stowage, and kept three of the party constantly employed with our long bamboo-handled dip-net, in fishing up specimens for the professor and his assistants. As the result of this we have a large number of fish eggs which we are watching in the process of hatching, many specimens of crustacea and of seaweed. The photographers, in the meanwhile, got themselves into readiness for real work by practicing incessantly upon us.

Thursday, we made Sambro light; soon pilot boat number one hailed us and put a man aboard, whom we neither needed nor wanted, and we were anchored off the market steps at Halifax. The run up the harbor was very pleasant. Bright skies, a fresh breeze off the land, and vessels all about us made many lively marine pictures. The rather unformidable appearing fortification, on account of which Halifax boasts herself the most strongly fortified city of America, together with the flag-ship Bellerophon and two other vessels of the Atlantic squadron, the Canada and the Thrush, the latter vessel until lately having been commanded by Prince George, gave the harbor and town a martial tone that was heightened upon our going ashore and seeing the red coats that throng the streets in the evening. Halifax, with its squat, smoky, irregular streets is well known, and its numerous public buildings, drill barracks, and well kept public gardens, all backed by the frowning citadel, probably need no description from me. After receiving the letters for which we came in, and sending the courteous United States Consul General, Mr. Frye, and his vice-consul, Mr. King, Colby '89, ashore with a series of college yells that rather startled the sleepy old town, we laid a course down the harbor, exchanged salutes with the steamship Caspian, and were soon ploughing along, before a fine south-west breeze for Cape Canso.

[Ward Room of the Julia Decker] While our little vessel is driving ahead with wind well over the quarter, groaning, as it were, at the even greater confusion in the wardroom than when we left Rockland, owing to the additional supplies purchased at Halifax, it may be well to briefly describe her appearance, when fitted to carry seventeen Bowdoin men in her hold in place of the lime and coal to which she has been accustomed. Descending, then, the forward hatch, protected by a plain hatch house, the visitor turns around and facing aft, looks down the two sides of the immense centreboard box that occupies the centre of our wardroom from floor to deck. Fastened to it are the mess tables, nearly always lighted by some four or five great lamps, which serve to warm as well, as the pile of stuff around and beneath the after-hatch house cuts off most of the light that would otherwise come down there. On the port side of the table runs the whole length of the box; two wooden settles serve for dining chairs and leave about four feet clear space next the "deacon's seat" that runs along in front of the five double-tiered berths. These are canvas-bottomed, fitted with racks, shelves, and the upper ones with slats overhead, in which to stow our overflowing traps.

At the after end, on both sides of the wardroom, are large lockers coming nearly to the edge of the hatch, in which most of the provisions are stowed. At the forward end, next to the bulkhead that separates us from the galley, are, on the port side, a completely equipped dark room in which many excellent pictures have already been brought to light, and on the starboard side a large rack holding our canned goods, ketchup, lime-juice, etc. Along the bulkhead are the fancy cracker boxes, tempting a man to take one every time he goes below, and under the racks are our kerosene and molasses barrels. Between the line of four double-tier berths on the starboard

4

side and the rack just described is a handy locker for oil clothes and heavy overcoats. Lockers run along under the lower berths, and trunks with a thousand other articles are stowed under the tables. A square hole cut in the bulkhead, just over the galley head, lets heat into the wardroom and assists the lamps in keeping us warm. As yet, in spite of some quite cold weather, we have been perfectly comfortable. Sometimes, however, odors come in as well as heat from the galley, and do not prove so agreeable. If to this description, clothes of various kinds, guns, game bags, boots, fishing tackle and books, should, by the imagination of the reader, to be scattered about, promiscuously hung, or laid in every conceivable nook and corner, a fair idea of our floating house could be obtained. On deck we are nearly as badly littered, though in more orderly fashion. Two nests of dories, a row boat, five water tanks, a gunning float, and an exploring boat, partly well fill the Julia's spacious decks. The other exploring boat hangs inside the schooner's yawl at the stern. Add to these two hatch houses, a small pile of lumber, and considerable fire wood snugly stowed between the casks, and you have a fair idea of our anything but clear decks. A yellow painted bust, presumably of our namesake Julia, at the end of figure-head, peers through the fog and leads us in the darkness; a white stripe relieves the blackness of our sides; a green rail surmounts all; and, backed by the forms of nineteen variously attired Bowdoin men, from professor, their tutor, alumnus, to freshmen, complete our description.

[The Fourth of July] Meanwhile the night, clear but windless, has come on, and we drift along the Nova Scotia coast, lying low and blue on our northern board. The Fourth dawns rather foggy, but it soon yields to the sun's rays and a good breeze which bowls us along toward the Cape. An elaborate celebration of the day is planned, but only the poem is finally rendered, due probably to increased sea which the brisk breeze raises incapacitating several of the actors for their assigned parts. The poem, by the late editor of '91's "BUGLE, " is worthy of preservation, but would hardly be understood unless our whole crowd were present to indicate by their roars the good points in it.

At night our constant follower, the fog, shuts in, and the captain steering off the Cape, we lay by, jumping and rolling in a northeast sea, waiting for daylight to assist us to Cape Canso Harbor and the Little Ant. About six next morning we form one of a fleet of five or six sail passing the striped lighthouse on Cranberry Island, and with a rush go through the narrow passage lined with rocks and crowded with fishermen. Out into the fog of Chedebucto Bay we soon pass and in the fog we remain, getting but a glimpse of the shore now and then, till we reach Port Hawkesbury.

JONA. P. CILLEY, JR.

* * * * *

ON BOARD THE "JULIA A. DECKER, "
OFF ST. JOHN'S BAY, NEWFOUNDLAND.

We are bowling along with a fine southwest wind, winged out, mainsail reefed and foresail two-reefed, and shall be in the straits in about two hours. The Julia is a flyer. Between 12 and 4 this morning we logged just 46 knots, namely, 13.5 miles per hour for four hours. I doubt if I ever went much faster in a sailing vessel. It is now about 10 o'clock, and we have made over 75 miles since 4.

All hands are on watch for a first glimpse of the Labrador coast, which will probably be Cape Armours with the light on it.

I wrote last time from Hawkesbury in the Gut of Canso. We laid there all day Monday, July 6th, as the wind, southeast in the harbor, was judged by everybody to be northeast out in George's Bay, and consequently dead ahead for us. Monday evening, at the invitation of the purser, we all went down aboard the "State of Indiana, " the regular steamer of the "State Line" between Charlottetown, P.E. I., and Boston, touching at Halifax, and in the Gut.

After going ashore we stayed on the wharf till she left, singing college songs, giving an impromptu athletic exhibition, etc., to the intense delight of about fifty small boys (I can't conceive where they all came from), and the two or three hundred servant girls going home to P. E.I. for a summer vacation.

I would put in here parenthetically, that since writing the above I have been on deck helping jibe the mainsail, as we have changed our course to about east by north, having rounded a couple of small low, sandy islands off the Bay of St. John, and now point straight into the strait of Belle Isle.

In the afternoon we examined some of the old red sandstone which underlies all that part of Cape Breton Island, found some good specimens, and some very plain and deep glacial scratches. There is also some coal and a good deal of shale in with the sandstone.

We had a good opportunity to see this, since the railroad connecting Port Hawkesbury with Sidney is new, having started running only last March, and hence the cuts furnished admirable fields in which to examine the geology. The road is surveyed and bed made along the

Cape Breton shore of the Gut nearly to the northern end, and when completed will be a delightful ride. I think the Gut for 10 miles north of Port Hawkesbury resembles the Hudson just by the Palisades. It is grander than Eggemoggin Reach and on a far larger scale than Somes' Sound. At the northern end it broadens and becomes just a magnificent waterway, without the grand scenery. We were becalmed nearly all day in George's Bay, at one time getting pretty near Antigonish, but got a breeze towards evening. We tried fishing several times but could not get a bite though several fishermen were in sight and trawls innumerable. We passed one fisherman, a fine three-master, just as we were coming out of the Gut from Frenchman's Bay, going home, but with very little fish.

I got the captain to call me about 4, Wednesday morning, to fish, but got none. We were then off North Cape, having had a good breeze all night. The wind was light all day, but towards the latter part of the afternoon commenced to blow from the southeast, kicking up a nasty sea very soon. We double reefed the mainsail reefed the foresail and hauled the flying jib down. About 8 P. M. we laid to with the jib hauled down, on the starboard tack. The wind had backed to the east about four points and was blowing a gale. About 12 M. it suddenly dropped, a flat calm, leaving a tremendous sea running from the southeast, combined with a smaller one from the east. Our motions, jumps, rolls and pitches, can be better imagined than described. It seemed at times that our bow and our stern were where the mastheads usually are, and our rails were frequently rolled under.

Rice and Hunt stood one watch, Cary and I the second, and here Rice, though a good sailor and an experienced yachtsman, finally succumbed. We hauled everything down with infinite difficulty, owing to the violent motion, and made it fast, then let her roll and pitch to her heart's content. A sorrier looking place than our wardroom, and a sicker set of fellows it would be hard to find. The dishes had some play in the racks, and kept up an infernal racket that I tried in every way to stop and could not. To cap all, the wind came off a gale northwest about 4 A. M., and made yet another sea. As soon as possible we set a double-reefed foresail, and then I turned in. When I turned out at noon we had made Newfoundland and set a whole foresail, jib and one reef out of the mainsail. We were becalmed, but found excellent fishing, so did not care. The sea had gone down and we began to enjoy the Norway-like rugged coast of Newfoundland. The mountains come right down to the water, and are about 1,400 feet high, by our measurement, using angular altitude by sextant and base line, our distance off shore as shown by our observation for latitude and longitude.

There are many deep, narrow-mouthed coves and harbors, a good number of islands and points making a most magnificent coast line. In many cases 50 or 75 fathoms are found right under the shore. Great patches of snow, miles in extent, cover the mountain sides. Great brown patches, which the professor thinks are washings from the fine examples of erosion, but which look to me like patches of brown grass as we see in Penobscot Bay on the islands, vary with what is apparently a scrubby evergreen growth and bald, bare rocks. As we are about 18 miles off, the blue haze over all makes an enlarged, roughened and much more deeply indented Camden mountain coast line. The bays are in some cases so deep that we can look into narrow entrances and see between great cliffs, only a few miles apart, a water horizon on the other side. We wished very much to get in towards the shore, but the calm and very strong westerly current, about 1-1/2 knots, prevented.

While enjoying the calm in pleasant contrast to our late shaking up, it will be well to introduce the members of the party whom Bowdoin has thought worthy to bear her name into regions seldom vexed by a college yell, and to whom she has entrusted the high duties of scientific investigation, in which, since the days of Professor Cleaveland, she has kept a worthy place.

[Members of the Expedition] In command is Prof. Leslie A. Lee, of the Biological Department of Bowdoin. With a life-long experience in all branches of natural history, the experience which a year in charge of the scientific staff of the U. S. Fish Commission Steamer "Albatross" in a voyage from Washington around Cape Horn to Alaska, and an intimate connection with the Commission of many year's standing, and the training that scholarly habits, platform lecturing and collegic instruction have given him, you see a man still young, for he was graduated from St. Lawrence University in 1872, and equal to all the fatigues that out-of-door, raw-material, scientific work demands.

The rest of the party have yet to prove their mettle, and of them but little can now be said. Dr. Parker, who, with the Professor, captain and mate, occupies the cabin proper, is an '86 man, cut out for a physician and thoroughly prepared to fulfil all the functions of a medical staff, from administering quinine to repairing broken limbs.

Cary of '87, who is even now planning for his struggle with the difficulties on the way to the Grand Falls, has had the most experience in work of the sort the expedition hopes to do, save the Professor and Cole. Logging and hunting in the Maine forests in the vicinity of his home in Machias, and fishing on the Georges from Cape Ann smacks, have fitted him physically, as taking the highest honors for scholarship at Bowdoin, teaching and university work in his chosen branch, have prepared him mentally, for the great task in which he leads.

Cole who accompanies him up Grand River, was Prof. Lee's assistant on the "Albatross, " and is well fitted by experience and by a vigorous participation in athletics at college before his graduation in '88.

From the expedition's actual starting place, Rockland, there are four members: Rice, the yachtsman, Simonton, Spear and the writer, all fair specimens of college boys, and eager to get some reflection from the credit which they hope to help the expedition to win.

Portland has two representatives: Rich, '92, and Baxter, 93, the latter our only freshman; while Bangor sends three: Hunt, '90, Hunt, '91, who has charge of the dredging, and Hastings the taxidermist.

W. R. Smith, another salutatorian of his class, is one of the many Maine boys whom Massachusetts has called in to help train the youth of our mother Commonwealth, and has been at the head of the High School at Leicester for the past year. He, too, is thought to equal in physical vigor his mental qualities, and has been selected to brave the hardships of the Grand River.

To complete the detail for this exploration, Young of Brunswick and of '92, has been selected, another athlete of the college, who has had, in addition to his training at Bowdoin, a year or more of instruction in the schools and gymnasiums of Germany.

Porter, Andrews, and Newbegin, the latter, the only man not from Maine, coming from Ohio, and only to be accounted for as a member of the expedition by the fact that his initials P. C. stand for Parker Cleaveland, finish the list, with but one exception and that is Lincoln.

The merry-maker and star on deck and below—except when the weather is too rough—he keeps the crowd good-natured when fogs, rain, head winds and general discomfort tend to discontent: and on shore he sees that the doctor is not too hard worked in making the botanical collections.

For two days we lazily drifted, the elements seeming to be making up for their late riot; but the weather was clear and bright, the scenery way off to our starboard was grand, and no one was troubled by the delay, except as the thoughts of the Grand River men turned to the great distance and the short time of their trip. At last, however, the breeze came, with which I opened this letter, and which we then hoped would continue till we reached Battle Harbor.

We just flew up the straits, saw many fishermen at anchor with their dories off at the trawls, schooners and dories both jumping in great shape; also a school of whales and an "ovea" or whale-killer, with a fin over three feet long sticking straight up. He also broke right alongside and blew. Considerable excitement attended our first sight of an iceberg; it was a rotten white one, but soon we saw a lot, some very dark and deep-colored.

[Red Bay] Our first sight of the long-desired coast was between Belle Armours Point and the cliffs near Red Bay, the thick haze making the outlines very indistinct. Just two weeks out from Rockland we made our first harbor on the Labrador coast. Red Bay is a beautiful little place, and with the added features of two magnificent icebergs close by which we passed in entering, the towering red cliffs on the left from which it takes its name, and the snug little island in the middle, and the odd houses we saw dotting the shores of the summer settlement of the natives, it seemed a sample fully equal to our expectations of what we should find in Labrador.

There is an inner harbor into which we could have gone, with seven fathoms of water and in which vessels sometimes winter as it is so secure, but we did not enter it because the captain was doubtful which of the two entrances to take and the chart seemed indefinite on the point. There are about one hundred and seventy-five people in the settlement, some of them staying there the year round, fishing in the summer and hunting the rest of the time. They have another settlement of winter houses at the head of the inner harbor, but, for convenience in getting at their cod traps, live on the island in the middle, and on the sides of the outer harbor in the summer. Their houses are made of logs about the size of small railroad ties, which are stood on end and clapboarded. The winter houses are built in a similar way with earth packed around and over them.

The party for Grand River—Cary, Cole, W.R. Smith and Young— have decided to dispense with a guide; very wisely, I think, from what I have seen of native Labradoreans. While the journey they undertake is one in which the skill of Indians or half-breeds, familiar with Labrador wildernesses would be of great value and would add to the comfort of our party, it is very doubtful if any living person has ever been to the falls or knows any more about the last, and probably the hardest part of the trip, than Cary. And, further, the travel is so difficult that about all a man can carry is supplies for himself; and the Indians cannot stand the pace that our men intend to strike; nor, if it should come to the last extremity, and a forlorn hope was needed to make a last desperate push for discovery or relief, could the Indian guides, so far as we have any knowledge of them, be relied on. That the boldest measures are often the surest, will probably again be demonstrated by our Grand River party.

We tried the exploring boats very thoroughly at Chateau Bay, three of us getting caught about six miles from the vessel in quite a blow, and the well-laden boat proved herself very seaworthy. When loaded, she still draws but little water, and is good in every way for the trip.

This letter was begun in the fine breeze off Newfoundland, but could not be mailed till the port of entry and post-office of Labrador, Battle Harbor, was reached. A week was consumed in getting from our first anchorage in Labrador to this harbor, as the captain was unaccustomed to icebergs, and properly decided to take no risks with them in the strong shifting currents and thick weather of the eastern end of the straits. The wind was ahead for several days, and the heavy squalls coming off the land in quick succession made us fear the wind would drop and leave us banging around in the fog that usually accompanies a calm spell, so we kept close to harbors and dodged in on the first provocation.

The season is three weeks late this year; the first mail boat has not yet arrived, though last year at this time she was on her second trip. The last report from the North—down the coast they call it—that went to Newfoundland and St. John's was "that it was impassable ice this side Hamilton Inlet. " A vessel—a steam sealing bark—

though, that was here yesterday and has gone to Sidney, C.B. I., reports now that the coast is clear to Hopedale. Beyond we know nothing about it.

On Henley and Castle Islands, at the mouth of Chateau Bay, are basaltic table-lands about half a mile across, perfectly flat on top and about two hundred feet high. We walked around one, went to its top and secured specimens from the columns. The famous "natural images" of men, are, to my eye, not nearly so good as the descriptions lead one to expect. The history of the place could hardly be guessed from its present barren, desolate, poverty-stricken appearance; but the remains of quite a fort on Barrier Point show some signs of former and now departed glory. It seems that it has been under the dominion of England, France and the United States, all of whom took forceful possession of it, and England and France have governed it. An American privateer once sacked the place, carrying away, I believe, about 3,500 pounds worth of property. Now, a very small population eke out a wretched existence by fishing, only a few remaining, living at the heads of the bays, in the winter, and most of them going home to Newfoundland.

The icebergs are in great plenty. I counted eighty from the basaltic table-land at one time, and the professor saw even more at once. Belle Isle is in plain sight from this place, looking like Monhegan from the Georges Islands, though possibly somewhat longer.

[Battle Harbor] Finally, as the wind showed no signs of changing, the captain, to our intense delight, decided to beat around to Battle Harbor and we anchored here at about 5:50 P. M., July 17th. Many of the icebergs we passed were glorious, and the scene was truly arctic. It was bitterly cold, and heavy coats were the order of the day. We passed Cape St. Charles, the proposed terminus of the Labrador Railroad to reduce the time of crossing the Atlantic to four days, saw the famous table-land, and soon opened Battle Harbor which we had to beat up, way round to the northward, to enter. It was slow business with a strong head current, but the fishermen say a vessel never came around more quickly. We found the harbor very small, with rocks not shown in chart or coast pilot, and had barely room to come to without going ashore. We went in under bare poles, and then had too much way on.

The agent for the Bayne, Johnston Co., which runs this place, keeping nearly all its three hundred inhabitants in debt to it, is a Mr.

Smith, who has taken the professor and seven or eight of the boys on his little steamer to the other side of the St. Lewis Sound. The doctor has gone with them to look after some grip patients, and the professor expects to measure some half-breed Eskimo living there. The boys are expecting to get some fine trout. The grip was brought to this region by the steamer bringing the first summer fishing colonies, and has spread to all and killed a great many.

There is an Episcopal rector here, Mr. Bull, who says everybody had it. I believe it is owing to his care and slight medical skill that none have died here. It is hard for this people to have such a sickness just as the fishing season is best. The doctor has opportunity to use all and far more than the amount of medicine he brought, much to Professor Lee's amusement. He is reaping a small harvest of furs, grateful tokens of his services, that many of his patients send him, and some of his presents have also improved our menu.

This place is named Battle Harbor from the conflict that took place here between the Indians and English settlers, aided by a man-of-war. The remains of the fight are now in a swamp covered with fishflakes. There are also some strange epitaphs in the village graveyard, with its painted wooden head-boards, and high fence to keep the dogs out. These latter are really dangerous, making it necessary to carry a stick if walking alone. Men have been killed by them, but last year the worst of the lot were exported across the bay, owing to a bold steal of a child by them and its being nearly eaten up. They are a mixture of Eskimo, Indian and wolf, with great white shaggy coats.

The steamer with mail and passengers from St. John's, Newfoundland, is expected every day, and as our rivals for the honor of rediscovering Grand Falls are probably on board, there is a race in store for us to see who will get to Rigolette first, and which party will start ahead on the perilous journey up the Grand River. As they have refused our offer of co-operation, we now feel no sympathy with their task, and will have but little for them till we see them, as we hope, starting up the river several days behind our hardy crew.

JONATHAN P. CILLEY, JR.

* * * * *

ON BOARD THE JULIA A. DECKER,
OFF BIRD ROCKS,
Gulf of St. Lawrence, Sept. 10, 1891.

While our little vessel is rushing through the blue waters of the gulf, apparently scorning the efforts of the swift little Halifax trader who promised to keep us company from the Straits to the Gut, and who, by dint of good luck and constant attention to sails has thus far kept her word, but is now steadily falling astern and to leeward, I will tell you about the snug little harbors, the bold headlands, barren slopes, and bird-covered rocks, and also the odorous fishing villages and the kind-hearted people with whom she has made us acquainted.

The Bowdoin scientific expedition to Labrador is now familiar with six of the seven wonders in this truly wonderful region. It has visited Grand Falls and "Bowdoin Canyon; " has been bitten by black flies and mosquitoes which only Labrador can produce, both in point of quality and quantity; has wandered through the carriage roads (!) and gardens of Northwest River and Hopedale; has dug over, mapped and photographed the prehistoric Eskimo settlements that line the shores, to the north of Hamilton Inlet; has made itself thoroughly conversant with the great fishing industry that has made Labrador so valuable, to Newfoundland in particular, and to the codfish consuming world in general; and finally is itself the sixth wonder, in that it has accomplished all it set out to do, though of course not all that would have been done had longer time, better weather and several other advantages been granted it.

It is almost another wonder, too, in the eyes of the Labradoreans, that we have, without pilot and yet without accident or trouble of any sort, made such a trip along their rocky coast, entered their most difficult harbors, and outsailed their fastest vessels, revenue cutters, traders and fishermen.

It will be a good many years before the visit of the "Yankee college boys, " the speed of the Yankee schooner and the skill and seamanship of the Yankee captain are forgotten "on the Labrador. "

The day after we left, July 19th, the mail steamer reached Battle Harbor with the first mail of the season. On board were Messrs. Bryant and Kenaston, anxiously looking for the Bowdoin party and estimating their chances of getting to the mouth of Grand River.

They brought with them an Adirondack boat, of canoe model, relying on the country to furnish another boat to carry the bulk of their provisions and a crew to man the same.

[Rigolette] When the news was received that we were a day ahead, the race began in earnest, the captain of the "Curlew" entering heartily into the sport and doing his best to overhaul the speedy Yankee schooner. When about half way up to Rigolette, on the third day from Battle Harbor, as we were drifting slowly out of "Seal Bight, " into which we had gone the previous night to escape the numerous icebergs that went grinding by, the black smoke, and later the spars of the mail steamer were seen over one of the numerous rocky little islets that block the entrance to the bight. The steamer's flag assured us that it was certainly the mail steamer, and many and anxious were the surmises as to whether our rivals were on board, and earnest were the prayers for a strong and favoring wind. It soon came, and we bowled along at a rattling pace, our spirits rising as we could see the steamer, in shore, gradually dropping astern. Towards night we neared Domino Run, and losing sight of the steamer, which turned out to make a stop at some wretched little hamlet that had been shut out from the outer world for nine months, at about the same time lost our breeze also. But the wind might rise again, and time was precious, so a bright lookout was kept for bergs, and we drifted on through the night. The next morning a fringe of islands shut our competitor from sight, but after an aggravating calm in the mouth of the inlet, we felt a breeze and rushed up towards Rigolette, only to meet the steamer coming out while we were yet several hours from that place.

Here we had our first experience with the immense deer-flies of Labrador. Off Mt. Gnat they came in swarms and for self-protection each man armed himself with a small wooden paddle and slapped at them right and left, on the deck, the rail, another fellow's back or head, in fact, wherever one was seen to alight. The man at the wheel was doubly busy, protecting himself, with the assistance of ready volunteers, from their lance-like bites, and steering the quickly moving vessel.

At last the white buildings and flag-staff which mark all the Hudson Bay Co. 's posts in Labrador, came in sight, snugly nestled in a little cove, beneath a high ridge lying just to the north-west of it, and soon we were at anchor. Our intention was to get into the cove, but the six knot current swept us by the mouth before the failing breeze enabled us to get in.

After supper the necessary formal call was made on the factor, Mr. Bell, by the professor, armed with a letter of introduction from the head of the company in London, and escorted by three or four of the party. A rather gruff reception, at first met with, became quite genial, when it appeared that we wanted no assistance save a pilot, and called only to cultivate the acquaintance of the most important official in Labrador.

With a promise to renew the acquaintance upon our return, we left, and after a hard pull and an exciting moment in getting the boat fast alongside, on account of the terrific current, we reached the deck and reported.

Our rivals were there, and had hired the only available boat and crew to transport them to North West River. This threw us back on our second plan, viz: to take our party right to the mouth of the Grand River ourselves, which involved a trip inland of one hundred miles to the head of Lake Melville. This it was decided to do, and after some delay in securing a pilot, owing to the transfer at the last moment of the affections of the first man we secured to the other party, John Blake came aboard and we started on our new experience in inland navigation. Just as we entered the narrows, after a stop at John's house to tell his wife where we were taking him, and to give her some medicine and advice from the doctor, we saw our rivals starting in the boat they had secured. That was the last we saw of them, till they reached North West River, two days after our party had started up the Grand River.

North West River is the name of the Hudson Bay Co. 's post at the mouth of the river of the same name, flowing into the western extremity of Lake Melville, about fifteen miles north of the mouth of Grand River. Hamilton Inlet proper extends about forty miles in from the Atlantic to the "Narrows, " a few miles beyond Rigolette, where Lake Melville begins. A narrow arm of the lake extends some unexplored distance east of the Narrows, south of and parallel to the southern shore of the inlet. The lake varies from five to forty miles in width and is ninety miles long, allowing room for an extended voyage in its capacious bosom. The water is fresh enough to drink at the upper end of the lake, and at the time of our visit was far pleasanter and less arctic for bathing than the water off any point of the Maine coast. About twenty miles from the Narrows a string of islands, rugged and barren, but beautiful for their very desolation, as is true of so much of Labrador, nearly block the way, but we found the channels deep and clear, and St. John's towering peak makes an excellent guide to the most direct passage.

One night was spent under way, floating quietly on the lake, so delightfully motionless after the restless movements of Atlantic seas. A calm and bright day following, during which the one pleasant swim in Labrador waters was taken by two of us, was varied by thunder squalls and ended in fog and drizzle, causing us to anchor off the abrupt break in the continuous ridge along the northern shore, made by the Muligatawney River. Although in an insecure and exposed anchorage, yet the fact that we were in an inclosed lake gave a sense of security to the less experienced, that the snug and rocky harbors to which we had become accustomed, usually failed to give on account of the roaring of the surf a few hundred yards away, on the other side of the narrow barrier that protected the rocky basin.

The following day was bright and showery by turns, but the heart's wish of our Grand River men was granted, and while the schooner lay off the shoals at the mouth of the river they were to make famous, they started as will be described, and the rest of the expedition turned towards North West River, hoping they, too, could now get down to their real work.

The noble little vessel was reluctant to leave any of her freight in so desolate a place, in such frail boats as the Rushtons seemed, and in the calm between the thunder squalls, several times turned towards them, as they energetically pushed up the river's mouth, and seemed to call them back as she heavily flapped her white sails. They kept steadily on, however, while the Julia, bowing to a power stronger than herself, and to a fresh puff from the rapidly rising thunder heads, speedily reached North West River.

North West River is a sportsman's paradise. Here we found the only real summer weather of the trip, the thermometer reaching 76 deg. F. on two days in succession, and thunder storms occurring regularly every afternoon. Our gunners and fishermen were tempted off on a long trip. One party planning to be away two or three days, but returning the following morning, reported tracks and sounds of large animals. They said the rain induced them to return so soon.

[Montagnais Indians] Here we found a camp of Montagnais Indians, bringing the winter's spoils of furs to trade at the post for flour and powder, and the other articles of civilization that they are slowly learning to use. They loaf on their supplies during the summer, hunting only enough to furnish themselves with meat, and then starve during the winter if game happens to be scarce. Measurements were made of some twenty-five of this branch of the Kree tribe, hitherto unknown to anthropometric science, and a full collection of household utensils peculiar to their tribe was procured. Several of the Nascopee tribe were with them, the two inter-marrying freely, and were also measured. The latter are not such magnificent specimens of physical development as the Montagnais, but their tribe is more numerous and seems, if anything, better adapted to thrive in Labrador than their more attractive brothers.

The only remains of their picturesque national costume that we saw, was the cap. The women wore a curious knot of hair, about the size of a small egg, over each ear, while the men wore their hair cut off straight around, a few inches above the shoulders.

In point of personal cleanliness, these people equal any aborigines we have seen, though their camp exhibited that supreme contempt for sanitation that characterizes every village except the Hudson Bay Co. 's posts on the Labrador coast, whether of Indians, Esquimaux or "planters, " as the white and half-breed settlers are called.

Some curious scenes were enacted while the professor was trading for his desired ethnological material. With inexhaustible patience and imperturbable countenance, he sat on a log, surrounded by yelping dogs, and by children and papooses of more or less tender ages and scanty raiment, playing on ten cent harmonicas that had for a time served as a staple of trade, struggling with the dogs and with their equally excited mothers and sisters for a sight of the wonderful basket from whose apparently inexhaustible depths came forth yet more harmonicas, sets of celluloid jewelry, knives, combs, fish-hooks, needles, etc., *ad infinitum*. The men, whose gravity equalled the delight of the women and children, held themselves somewhat aloof, seldom deigning to enter the circle about the magic basket, and making their trades in a very dignified and careless fashion.

That these people are capable of civilization there can be no doubt. Missing the interpreter, without whom nothing could be done, the professor inquired for him and learned that he had returned to his wigwam. Upon being summoned he said he was tired of talking. Thereupon the professor bethought himself and asked him if he wanted more pay. The interpreter, no longer tired, was willing to talk all night.

The camp was in a bend of the river and at the head of rapids about four miles from the mouth, up which we had to track, that is, one man had to haul the boat along by the bank with a small rope called a tracking line, while another kept her off the rocks by pushing against her with an oar. At that point the river opened out into a beautiful lake from one to two miles in width, whose further end we could not see. As this river never has been explored to its head, we were surprised that Messrs. Bryant and Kenaston, who were ready for their inland trip about a week after our party had started up the Grand River, had not chosen it as a field for their work rather than follow in the footsteps of our expedition.

[A carriage road] Of all Labrador north of the Straits, North West River alone boasts a carriage road. To be sure, there are neither horses nor carriages at that post, but when Sir Donald A. Smith, at present at the head of the Hudson Bay Co. 's interests in Canada, but then plain Mr. Smith, factor, was in charge of that post his energy made the place a garden in the wilderness, and in addition to luxuries of an edible sort, he added drives in a carriage through forest and by shore, for about two miles, on a well made road. Now, we are informed there is not a horse or cow north of Belle Isle. The present factor, Mr. McLaren, is a shrewd Scotchman, genial and warm-hearted beneath a rather forbidding exterior, as all of our party who experienced his hospitality can testify.

In spite of all its attractions we could not stay at North West River. In five weeks we were to meet our river detail at Rigolette, and during that time a trip north of 400 miles was to be made and the bulk of the expedition's scientific work to be done.

Our day's sail, with fresh breezes and favoring squalls, took us the whole length of the delightful lake, whose waters had seldom been vexed by a keel as long as the Julia's, and brought us to an anchor off Eskimo Island. Here we had one of our regular fights with the mosquitoes, the engagement perhaps being a trifle hotter than usual, for they swarmed down the companion way every time the "mosquito door, " of netting on a light frame hinged to the hatch house, was opened, in brigades and divisions and finally by whole army corps, till we were forced to retreat to our bunks, drive out the intruding hosts, which paid no respect whatever to our limited 6x3x3 private apartments, by energetically waving and slapping a towel around, then quickly shutting the door of netting, also on a tightly fitting frame, and devoting an hour or two at our leisure to demolishing the few stragglers that remained within; or possibly the whole night, if an unknown breach had been found by the wily mosquito somewhere in our carefully made defenses. A few bones were taken from the Eskimo graves that abound on the island, but the mosquitoes seriously interfered with such work and the party soon returned to the vessel. The absolutely calm night allowed the mosquitoes to reach us and stay; and in spite of its brevity and the utter stillness of the vast solitude about us, broken only now and then by a noise from the little Halifax trader whose acquaintance we here made for the first time, and of whom we saw so much on our return voyage across the gulf, or by the howling of wolves and Eskimo dogs in the distance, we were glad when it was over and a morning breeze chased from our decks the invading hosts.

A short stop at Rigolette, to send about fifty letters ashore, a two days' delay in a cold, easterly storm at Turner Cove, on the south side of the inlet, when the icy winds, in contrast to the warm weather we had lately enjoyed, made us put on our heavy clothes and, even then, shiver—a delay, however, that we did not grudge, for we were in a land of fish, game and labradorite—this of a poor quality, as we afterward learned—and where the doctor had more patients than he could easily attend to. At last a pleasant Sunday's run to Indian Harbor got us clear of Hamilton Inlet. There we found the usual complement of fish and fishing apparatus, but with the addition of a few Yankee vessels and a church service.

The latter we were quite surprised to find, and several went, out of curiosity, and had the satisfaction of finding a small room, packed with about fifty human beings, with no ventilation whatever, and of sitting on seats about four inches wide with no backs. The people were earnest and respectful, but did not seem to understand all that was said, as, perhaps, is not to be wondered at, since they are the poorest class of Newfoundlanders.

Indian Harbor is like so many others on the coast, merely a "tickle" with three ticklish entrances full of sunken rocks and treacherous currents. The small islands that make the harbor are simply bare ledges, very rough and irregular in outline. The fishing village, also, like all others, consists of little earthen-covered hovels, stuck down wherever a decently level spot fifteen feet square can be found, and of fishing stages running out from every little point and cove, in which the catch is placed to be taken care of, and alongside of which the heavy boats can lie without danger of being smashed by the undertow that is continually heaving against the shore.

[Storm and fog] A two days' run brought us up to Cape Harrigan, rounding which we went into Webeck Harbor, little thinking that in that dreary place storm and fog would hold us prisoners for five days. That was our fate, and even now we wonder how we lived through that dismal time.

One day served to make us familiar with the flora, fauna, geography and geology of the region, for it was not an interesting place from a scientific point of view, however the fishermen may regard it, and after the departure of the mail steamer, leaving us all disappointed in regard to mail, time dragged on us terribly.

Two or three of the more venturesome ones could get a little sport by pulling a long four miles down to the extremity of Cape Harrigan, where sea pigeon had a home in the face of a magnificent cliff, against the bottom of which the gunners had to risk being thrown by the heavy swell rolling against it, as they shot from a boat bobbing like a cork, at "guillemots" flying like bullets from a gun out of the face of the cliff. One evening a relief party was sent off for two who had gone off to land on a bad lee shore and were some hours overdue. To be sure the missing ones arrived very soon, all right, while the search party got back considerably later, drenched with spray and with their boat half full of water, but the incident gave some relief from the monotony.

Another evening several visiting captains and a few friends from ashore were treated to a concert by the Bowdoin Glee and Minstrel Club. All the old favorites of from ten years ago and less were served up in a sort of composite hash, greatly to the delight of both audience and singers.

[Abundance of codfish] At Webeck Harbor, which we came to pronounce "Wayback, " probably because it seemed such a long way back to anything worthy of human interest, we saw the business of catching cod at its best. They had just "struck a spurt, " the fishermen said, and day after day simply went to their traps, filled their boats and bags, took the catch home, where the boys and "ship girls" took charge of it, and returned to the traps to repeat the process. An idea of the amount of fish taken may be given by the figures of the catch of five men from one schooner, who took one thousand quintals of codfish in thirteen days. We obtained a better idea of the vast catch by the experience of one of our parties who spent part of a day at the traps, as the arrangement of nets along the shore is called, into which the cod swim and out of which they are too foolish to go. They are on much the same plan as salmon weirs, only larger, opening both ways, and being placed usually in over ten fathoms of water and kept in place by anchors, shore lines, and floats and sinkers. Once down they are usually kept in place a whole season. The party were in a boat, inside the line of floats, so interested in watching the fishermen making the "haul, " as the process of overhauling the net and passing it under the boat is called, by which the fish are crowded up into one corner where they can be scooped out by the dozen, that they did not notice that the enormous catch was being brought to the surface directly under them till their own boat began to rise out of the water, actually being grounded on the immense shoal of codfish.

It was a strange sensation and makes a strange story. All the time that we were storm-stayed at Webeck the "spurt" continued, and the trap owners were tired but jubilant. The "hand-lining" crews were correspondingly depressed, for, though so plenty, not a cod would bite a hook. It is this reason, that is, because an abundance of food brings the cod to the shores in great numbers and at the same time prevents them from being hungry, that led to the abandonment of trawling and the universal adoption of the trap method. We did not see a single trawl on the coast, and it is doubtful if there was one there in use.

During these spurts, the day's work just begins, in fact, after the hard labor of rowing the heavy boats out, perhaps two miles, to the trap, hauling, mending the net, loading and unloading the fish—always a hard task and sometimes a very difficult one on account of the heavy sea—has been repeated three or four times; for the number of fish is so great that the stage becomes overloaded by night, and the boat crews then have to turn to and help take care of the catch and clear the stage for the next day's operations. Till long after midnight the work goes merrily on in the huts or shelters over the stages, for the hard work then means no starvation next winter in the

Newfoundland homes, and the fish are split, cleaned, headed, salted and packed with incredible rapidity.

The tired crews get an hour or two of sleep just as they are; then, after a pot of black tea and a handful of bread, start out to begin the next day's work, resting and eating during the hour between the trips, and then going out again, and repeating the some monotonous round over and over till we wondered how they lived through it, and what was to be done with all the fish. When there is a good breeze the boats are rigged and a large part of the weary labor of rowing is escaped. How tired the crews would look as the big twenty-four feet boats went dashing by our vessel in the fog and rain, on the outward trip, and how happy, though if possible more tired, as they came back three or four hours later, loaded to the gunwale with cod, and thinking, perhaps, of the bags full that they had left buoyed near the trap because the boat would not carry the whole catch. It is a hard life, and no wonder the men are not much more than animals; but they work with dogged persistence, for in a little more than two months enough must be earned to support their families for the year. When the "spurt" ends the crews get a much needed rest, and attend to getting a supply of salt ashore from the salt vessel from Cadiz, Spain, one of which we found lying in nearly every fishing harbor, serving as a storehouse for that article so necessary to the fishermen.

As to the magnitude of the industry, it is estimated that there are about 3,000 vessels and 20,000 men employed in it during the season. Some of the vessels are employed in merely bringing salt and taking away the fish, notably the great iron tramp steamers of from 1,500 to 2,000 tons, which seem so much out of place moored to the sides of some of the little rocky harbors. The average catch in a good year is, we were informed, from four to six hundred quintals in a vessel of perhaps forty tons, by a crew of from four to eight men. The trap outfit costs about $500 and is furnished by the large fish firms in Newfoundland, to be paid for with fish. As the market price, to the fishermen, is from five dollars to six dollars a quintal, the value of the industry is at once apparent.

The great bulk of the fish go to Mediterranean ports direct, to Catholic countries, chiefly, and also to Brazil. The small size and imperfect curing which the Labrador summer allows make the fish almost unsalable in English and American markets. Many of the cod are of the black, Greenland variety, which are far less palatable, and are usually thrown away or cured separately for the cheaper market.

All storms come to an end finally, and at last the sun shone, the windlass clanked and we were underway. The long delay seemed to have broken our little schooner's spirits, for after being out three or four hours we had gone but as many miles, and those in the wrong direction.

At length the gentle breeze seemed to revive her and we gently slipped by the Ragged Islands and Cape Mokkavik. That Sunday evening will long be remembered by us, for in addition to the delight we felt at again moving northward, and the charm of a bright evening with a gentle, fair wind and smooth water, allowing us to glide by hundreds of fulmar and shearwater sitting on the water, scarcely disturbed by our passage, the moon was paled by the brightest exhibition of the aurora we saw while in northern waters. Its sudden darts into new quarters of the heavens, its tumultuous waves and gentle undulations, now looking like a fleecy cloud, now like a gigantic curtain shaken by still more gigantic hands into ponderous folds—all were reflected in the quiet water and from the numerous bergs, great and small, that dotted the surface, till the beholder was at times awe-struck and silent, utterly unable to find words with which to express himself.

The next day we rounded Gull Island, which we identified with some difficulty, owing to the absence of the flagstaff by which the coast pilot says it can be distinguished, and, after a delightful sail up the clear sound leading through the fringe of islands to Hopedale, we spied the red-roofed houses and earth-covered huts, the mission houses and Eskimo village, of which the settlement consists, snugly hidden behind little "Anatokavit, " or little Snow Hill Island, at the foot of a steep and lofty hill surmounted by the mission flagstaff. Here we were destined to pass five days as pleasant as the five at Webeck had been tedious.

[Hopedale] The harbor at Hopedale is the best one we visited on the coast. The twelve miles of sound, fringed and studded with islands, completely broke the undertow which had kept our vessel constantly rolling, when at anchor, in every harbor except those up Hamilton Inlet and Lake Melville.

About two miles south of us a vast, unexplored bay ran for a long distance inland, while to the north, looking from Flagstaff Peak, we could see Cape Harrigan and the shoals about it, the numberless inlets, coves and bays which fill in the sixty miles to Nain. We were very much disappointed at our inability to go north to that place, but before our start from the United States Hopedale had been named as the point with which we would be content if ice and winds allowed us to reach it, and that point proved the northern limit of our voyage.

About half a mile across the point of land on which the missionary settlement lies, is the site of the pre-historic village of "Avatoke, " which means "may-we-have-seals. " It consisted of three approximately circular houses, in line parallel with the shore, at the head of a slight cove, backed to the west by a high hill, and with a fine beach in front, now raised considerably from the sea level. Along the front of the row of houses were immense shell heaps, from which we dug ivory, that is, walrus teeth; carvings, stone lamps, spear heads, portions of kyaks, whips, komatiks, as the sleds are called, etc., etc., and bones innumerable of all the varieties of birds, fish and game on which the early Eskimo dined; as well as remnants of all the implements which Eskimos used in the household generations ago, and which can nearly all now be recognized by the almost identically shaped and made implements in the houses of Eskimos there in Hopedale, so little do they change in the course of centuries. The village has been completely deserted for over one hundred years, and was in its prime centuries before that, so the tales of its greatness are only dim Eskimo traditions.

The houses were found to average about thirty-five feet across on the inside; are separated by a space of about fifteen feet, and each had a long, narrow doorway or entrance, being almost exactly in line. The walls are about fifteen feet thick and now about five feet high, of earth, with the gravel beach for a foundation. The inside of the wall was apparently lined with something resembling a wooden bench. When, in one of the houses, the remains of the dirt and stone roof that had long since crushed down the rotten poles and seal skins that made the framework and first covering, had been carefully removed, the floor was found to be laid with flagstones, many three or four feet across, closely fitted at the edges and well laid in the gravel so as to make a smooth, even floor. This extended to the remains of the bench at the sides, and made a dwelling which for Eskimo land must have been palatial. The evidences of fire showed the hearth to have been near the center of the floor, a little towards the entrance, in order to get the most from its heat. The Hopedale Eskimo were themselves surprised at the stone floor, but one old man remembered that he had been told that such floors were used long ago, in the *palmier* days of Eskimo history, if such an expression is fitting for an arctic people.

A village arranged on a similar plan, except that the houses were joined together, was found to constitute the supposed remains of a settlement on Eskimo Island in Lake Melville.

In both cases the front of the row is towards the east, and the houses are dug down to sand on the inside, making their floors somewhat below the level of the ground.

[Eskimos] A more thorough investigation than we were able to make of the remains at Eskimo Island would undoubtedly yield much of interest and value, for they were if anything even older than those at Hopedale, probably having been abandoned after the battle between Eskimo and Indians, fought on the same island, which has now become a tradition among the people.

Five days were spent in this most interesting ethnological work, and hard days they were, too, as well as interesting, for the mosquitoes, black flies and midges were always with us; but on the other hand, the Eskimo interpreter was continually describing some national custom which some find would suggest to him, and very ingenious he proved to be in naming finds which we were entirely ignorant of or unable to identify.

The race as a whole is exceedingly ingenious, quick to learn, handy with tools, and also ready at mastering musical instruments. One of the best carpenters on the Labrador is an Eskimo at Aillik, from whom we bought a kyak; and at Hopedale in the winter they have a very fair brass band. The art of fine carving, however, seems to be dying out among them, and now there is but one family, at Nain, who do anything of the sort worthy the name of carving. Prof. Lee obtained several very fine specimens for the Bowdoin cabinets, but as a rule it is very high priced and rare. Most of it is taken to London by the Moravian mission ship, and has found its way into English and Continental museums. The figures of dogs, of Eskimos themselves, as well as of kyaks and komatiks, seals, walrus, arctic birds and the like are most exquisitely done.

The mission itself deserves a brief description. It was founded in 1782 and has been steadily maintained by the Moravian society for the furtherance of the Gospel, and is now nearly self-supporting. There are three missions of the society in Labrador, the one at Nain being the chief and the residence of the director, but Hopedale is very important as it is the place where the debasing influence of the traders and fishermen is most felt by the Eskimo, and the work of the missionaries consequently made least welcome to them. However, they have persevered, in the German fashion, and seem to have a firm hold on the childlike people which the seductions of the traders cannot shake off.

There are five missionaries now stationed at Hopedale: Mr. Townly, an Englishman, whose work is among the "planters" and fishermen; Mr. Hansen, the pastor of the Eskimo church; and Mr. Kaestner, the head of the mission, and in special charge of the store and trading, by which the mission is made nearly self-supporting; Mrs. Kaestner and Mrs. Hansen complete the number, and the five make up a community almost entirely isolated from white people during nine months of every year.

The fact that the two ladies spoke very little English was somewhat of a drawback, but detracted very slightly from our enjoyment of Mrs. Hanson's delightful singing and none at all from our appreciation of her playing on the piano and organ. To get such a musical treat in the Labrador wilds was most unexpected and for that reason all the more thoroughly enjoyed.

The mission house is a yellow, barn-like building, heavily built to prevent its being blown away, snugly stowed beneath a hill, and seeming like a mother round which the huts of the Eskimo cluster. The rooms in which we were so pleasantly entertained were very comfortably and tastily furnished, a grand piano in one of them seeming out of place in a village of Labrador, but so entirely in harmony with its immediate surroundings that we hardly thought of the strangeness of it, within a few yards of a village of pure Eskimo, living in all their primitive customs and in their own land.

A few rods behind the mission are the gardens, cut up into small squares by strong board fences to prevent the soil from blowing away, each with a tarpaulin near by to spread over it at night. In this laborious way potatoes, cabbages and turnips are raised. In a large hothouse the missionaries raise tomatoes, lettuce, and also flowers, but for everything else, except fish, game and ice, they have to depend on the yearly visit of the Moravian mission ship. She left for Nain just the day before we reached Hopedale, and after unloading supplies, etc., there, she proceeds north, collecting furs and fish until loaded, and then goes to London.

About fifty Eskimos were measured and collections made of their clothing, implements of war and chase and household utensils, which are the best of our collections, for the World's Fair and the Bowdoin museums.

After spending these five pleasant and profitable days at Hopedale, and regretfully looking out by Cape Harrigan, to Nain, whose gardens are the seventh wonder of Labrador, through which, reports say, one can walk for two miles, and whose missionaries, warned of our coming, were making ready to give us a warm reception; and near it Paul's Island, on which was so much of interest to our party; all this we thought of mournfully as our vessel's head was pointed southward and we sped along, reluctant on this account, and yet eager to hear of the success of our boldest undertaking, the Grand River exploration party.

At Aillik, where there is an abandoned Hudson Bay Co. 's post, we measured a few more Eskimo, obtained a kyak, which a day or two later nearly became a coffin to one of our party, and tried a trout stream that proved the best we found in Labrador. In about an hour, three of our party caught over eighty magnificent trout, and, naturally, returned much elated.

The next day we poked the Julia's inquisitive nose into one or two so-called but misnamed harbors that afforded very little shelter, and had a threatening and deserted look which, although the characteristic of the Labrador shore in general, has never been noticeable in the harbors we have visited. Many of them are very small, and in some it is necessary to lay quite close to the rocks, but yet we have had no trouble from the extremely deep water that we were told we should have to anchor in, nor yet from getting into harbors so small that it was hard to get out of them.

[Tickles] As a matter of fact, experience has taught the fishermen to use "tickles, " as narrow passages are called, for harbors, that there may always be a windward and a leeward entrance. In a few cases where the harbor is too small to beat out of, and has no leeward entrance, we have found heavy ring bolts fastened into proper places in the cliffs, to which vessels can make their lines fast, and warp themselves into weatherly position from which a course can be laid out of the harbor.

Meanwhile we are again approaching the Ragged Islands, which we passed just as we were beginning that memorable Sunday evening sail, about fifteen miles from the place we so much dread, Webeck Harbor.

On them we found the only gravel bed we saw in Labrador, and yet their name is due to the rough piled basaltic appearing rock, that proved on close examination to be much weathered sienite and granite. The harbor is an open place amidst a cluster of rocky islets, and we found it literally packed with fishing vessels. Here an afternoon was spent making pictures and examining the geology of these interesting islands, and here the adventure of the kyak, before referred to, took place.

Our fur trader thought he would take a paddle, but had not gone three lengths before he found that he was more expert in dealing with Eskimo furs than in handling Eskimo boats. He rolled over, was soon pulled alongside, and clearing himself from the kyak climbed aboard, just as our gallant mate, his rescuer, rolled out of his dory into the water and took a swim on his own account. All hands were nearly exploded with laughter as he rolled himself neatly into the dory again and climbed aboard, remarking, "That's the way to climb into a dory without capsizing her, " as he ruefully shook himself. We wanted to ask him if that was the only way to get out of a dory without turning her over, but we forebore.

The next morning as we got clear of the harbor, a trim looking schooner of our size was sighted just off Cape Harrigan, about ten miles ahead. The breeze freshening we gradually overhauled her, and finally, while beating into Holton harbor, one of the most dangerous entrances on the coast, by the way, we passed her, and noticing her neat rig and appearance guessed rightly we had beaten the representatives of the Newfoundland law and the collector of her revenues from this coast.

Mr. Burgess, who combines in one unassuming personage the tax and customs collector, the magistrate and the commissioner of poor relief from Labrador, afterward told us that the "Rose" had been on the coast for thirteen years and had been outsailed for the first time. The next morning we again beat her badly, in working up to Indian Harbor, and only then would he acknowledge himself fairly beaten.

[Puffins and Auks] Saturday, the 22d of August, having yet three days before we were due at Rigolette to meet our Grand River party, we made memorable in the annals of the puffins and auks of the Heron Islands by spending three or four hours there and taking aboard three hundred and seventy-eight of them. Many more of them were killed but dropped into inaccessible places or into the water and could not be saved.

The sound of the fusilade from over twenty gunners must have resembled a small battle, but it did not drive the birds away, and as we left they seemed thicker than ever. Not only was the air alive with them, but as one walked along the cliffs they would dart swiftly out of holes in the rocks or crevices, so the earth, too, seemed full of them. It was great sport for a time, but soon seemed too much like slaughter, and we would let the awkward puffins, with their foolish eyes and Roman noses, come blundering along within a few feet of our muzzles, and chose rather the graceful, swift motioned auks and guillemots, whose rapid flight made them far more sportsmanlike game.

The next day, though Sunday, had to be spent in taking care of the best specimens, and the game was not fully disposed of for several days. Our bill of fare was correspondingly improved for a few days.

Three days were consumed in beating up to Rigolette. At Indian Harbor we had heard rumors of the return of some party from Grand River on account of injuries received by one of the men, but the description applied best to the second party, and we decided it must refer to Bryant or Kenaston. Near Turner's Cove we found more rumors, but nothing definite enough to satisfy our growing anxiety, and at last, unable to bear the suspense any longer, three of the party took a boat and started to row the fifteen miles between us and Rigolette, while the vessel waited for a change of tide and a breeze.

Alternate hope and fear lent strength to our arms as we drove the light boat along, and soon we came in sight of the wharf. There we saw a ragged looking individual, smoking a very short and black clay pipe, with one arm in a sling, who seemed to recognize us, and waved his hat vigorously with his well arm. Soon we recognized Young and were pumping away at his well hand in our delight at finding his injuries no worse, and that Cary and Cole were yet pushing on, determined to accomplish their object.

Young's hand had been in a critical state; the slight injury first received unconsciously, from exposure and lack of attention had caused a swelling of his hand and arm that was both extremely painful and dangerous, and which, the doctor said, would have caused the loss of the thumb, or possibly of the whole hand, had it gone uncared for much longer. Of course it was impossible to leave a man in such a condition, or to send him back alone. So Smith very regretfully volunteered to turn back—at a point where a few days more were expected to give a sight of the Falls, and when all thought the hardest work of the Grand River party had been accomplished—and accompany Young back to Rigolette.

It was a great sacrifice of Smith's personal desires, to be one of the re-discoverers of the falls, to the interests of the expedition, and it involved a great deal of hard work, for, after paddling and rowing all day, he had to build and break camp every night and morning, as Young's hand grew steadily worse and was all he could attend to. At the mouth of the river, which was reached in shorter time than was expected, and without accident, Young obtained some relief from applications of spruce gum to his hand by Joe Michelini, a trapper and hunter, famous for his skill in all Labrador. Northwest River was reached the following day, and after a few days of rest for Smith, during which time Young's injury began to mend also under the influences of rest and shelter, they hired a small schooner boat to take them to Rigolette. On the passage they were struck by a squall in the night, nearly swamped, and compelled to cut the Rushton boat adrift in order to save themselves. The next day they searched the leeward shore of the lake in vain, and had to go on without her, arriving at Rigolette without further accident, and had been there about a week when we arrived. The boat was picked up later in a badly damaged condition, and given to the finder.

While Young outlined his experience we hunted up Smith, who had been making himself useful as a clerk to the factor at the Post, Mr. Bell, and all went on board the Julia as soon as she arrived, to report and relieve in a measure the anxiety of the professor and the boys.

[Anxious waiting] The day appointed for meeting the river party was the day on which we reached Rigolette, August 25th, and so a sharp lookout was kept for the two remaining members of the party, on whom, now, the failure or success of that part of the expedition rested. As they did not appear, we moved up to a cove near Eskimo Island, at the eastern end of Lake Melville, the following day, and there spent four days of anxious waiting. Some dredging and geological work was done, and an attempt was made to examine more carefully the remains of the Eskimo village before referred to on Eskimo Island, which some investigators had thought the remains of a Norse settlement. The turf was too tough to break through without a plow, and we had to give it up, doing just enough to satisfy ourselves that the remains were purely Eskimo.

All the work attempted was done in a half-hearted manner, for our thoughts were with Cary and Cole, and as the days went by and they did not appear, but were more and more overdue, our suspense became almost unbearable. Added to this was the thought that we could wait but a few days more at the longest, without running the danger of being imprisoned all winter, and for that we were poorly prepared.

The first day of September we moved back to Rigolette to get supplies and make preparations for our voyage home, as it was positively unsafe to remain any longer. The Gulf of St. Lawrence is an ugly place to cross at any time in September, for in that month the chances are rather against a small vessel's getting across safely.

It was decided that the expedition must start home on Wednesday, the 2nd, and that a relief party should be left for Cary and Cole. With heavy hearts the final preparations were made, and many were the looks cast at the narrows where they would be seen, were they to heave in sight.

At last, about 3.30 p. m. Tuesday, the lookout yelled, "Sail ho! in the narrows, " and we all jumped for the rigging. They had come, almost at the last hour of our waiting, and with a feeling of relief such as we shall seldom again experience we welcomed them aboard and heard their story.

* * * * *

ON BOARD THE JULIA A. DECKER,
GUT OF CANSO.

Bowdoin pluck has overcome Bowdoin luck, and though they literally had to pass through fire and water, the Bowdoin men, from the Bowdoin College Scientific Expedition to Labrador have done what Oxford failed to do, and what was declared well nigh impossible by those best acquainted with the circumstances and presumably best judges of the matter. Austin Cary and Dennis Cole, Bowdoin '87 and '88, respectively, have proven themselves worthy to be ranked as explorers, and have demonstrated anew that energy and endurance are not wanting in college graduates of this generation.

A trip up a large and swift river, totally unknown to maps in its upper portions, for three hundred miles, equal to the distance from Brunswick, Me., to New York City, in open fifteen feet boats, is of itself an achievement worthy of remark. But when to this is added the discovery of Bowdoin Canon, one of the most remarkable features of North America, the settlement of the mystery of the Grand Falls, and the bringing to light of a navigable waterway extending for an unbroken ninety miles, and three hundred miles in the interior of an hitherto unknown country, something more than remark is merited.

July 26th the schooner hove to about four miles from the mouth of the Grand River, the shoals rendering a nearer approach dangerous, and the boats of the river detachment were sent over the side, taken in tow by the yawl, and the start made on what proved the most eventful part of the Labrador expedition. Cheers and good wishes followed the three boats till out of hearing, and then the Julia gathered way and headed for North West River, while the party in the yawl with the two Rushtons in tow put forth their best efforts to reach the mouth of the river and a lee before the approaching squall should strike them.

The squall came first, and as it blew heavily directly out of the river, we could simply lay to and wait for it to blow over. Then a calm followed and by the time the next squall struck we were in a comparative lee. After the heaviest of it had passed, the Grand River boys clambered into their boats and with a hearty "good by" pulled away for the opening close at hand. The yawl meantime had grounded on one of the shoals, but pushing off and carefully dodging the boulders that dot those shallow waters, she squared away for North West River, following around the shore, and with the aid of a fresh breeze reached the schooner shortly after 10 o'clock P. M.

[Grand River] The river party was made up of Austin Cary in charge, and W. R. Smith, '90, occupying one boat, and Dennis Cole and E. B. Young, '92, with the other, all strong, rugged fellows, more or less acquainted with boating in rapid water, and well equipped for all emergencies. Their outfit included provisions for five weeks, flour, meal, buckwheat flour, rice, coffee, tea, sugar, beef extract, tins of pea soup, beef tongue, and preserves. They were provided with revolvers, a shot gun and a rifle, and sufficient ammunition, intending to eke out the stores with whatever game came in their way, although the amount of time given them would not allow much hunting. All the supplies, including the surveying, measuring and meteorological instruments, were either in tins or in water-tight wrappings, while the bedding and clothing were protected by rubber blankets. The boats, made by Rushton, the Adirondack boat-builder, were of cedar, fifteen feet long, five feet wide, double-ended, and weighed eighty pounds apiece. A short deck at each end of the boats covered copper air-tanks, which made life-boats of them and added much to their safety. Each boat was equipped with a pair of oars, a paddle and about one hundred feet of small line for tracking purposes. Proceeding about three miles the first camp was made on the south shore of Goose Bay, amid an abundance of mosquitoes. The next day twenty-five miles were made through shoals that nearly close the river's mouth, leaving but one good channel through which the water flows very swiftly, by the house of Joe Michelin, the trapper, at which six weeks later two very gaunt and much used up men were most hospitably received. Here another night was spent almost without sleep, owing to the mosquitoes.

Tuesday a large Indian camp was passed, the big "pool, " at the foot of the first falls and some three miles long, rowed across, and at noon the carry was begun. It was necessary to make seventeen trips and four and one half hours were used in the task. When the last load had been deposited at the upper end of the carry, the men threw themselves down on the bank utterly weary, and owing to the loss of sleep the two previous nights, were soon all sound asleep. In consequence camp was made here, and the first comfortable night of the trip passed. Including the carry eight miles was the day's advance.

The twenty-five miles of the next day were made rowing and tracking up the Porcupine rapids through a series of small lakes, one with a little island in the centre deceiving our boys for awhile into thinking they had reached Gull Island Lake, and then up another short rapid at the head of which the party encamped.

Sixteen miles were made next day by alternate rowing and tracking, the foot of Gull Island Lake was reached, and after dinner it was crossed in one and a half hours. Then the heaviest work of the trip thus far was struck and camp was made, about half way up Gull Lake rapid. Supper was made off a goose shot the previous day. It was necessary to double the crews in getting up the latter part of Gull Island rapids, and finally a short carry was made just at noon to get clear of them. From the fact that the light, beautifully modelled boats required four men to take them up the rapids we may get some idea of the swiftness of the river as well as the difficulties attending the mode of travelling. As the river in its swiftest parts is never less than half a mile wide, and averages a mile, it can readily be seen that it is a grand waterway, well deserving its name.

Nine miles were made this day and camp was reached at the beginning of rough water on the Horse Shoe Rapid. Here the first evidence of shoes giving out was seen. Constant use over rough rocks while wet proved too much for even the strongest shoes, and when Cary and Cole returned there was not leather enough between them to make one decent shoe. Rain made the night uncomfortable, as the light shelter tent let the water through very easily and was then of little use. At other times the tents were very comfortable. Upon arriving at the spot selected two men would at once set about preparing the brush for beds, pitching the tent, etc., while the other provided wood for the camp and for the cook, in which capacity Cary officiated. I cannot do better than use Cary's own words in reference to his "humble but essential ministrations. " "Camp cooking at best is rather a wearing process, but the agonies of a man whose hands are tangled up in dough and whom the flies becloud, competing for standing room on every exposed portion of his body, can be imagined only by the experienced. "

The party believed that a good night's rest was indispensible where the day was filled with the hardest kind of labor, and spared no pains to secure them. Even on the return Cary and Cole, when half starved, stuck to their practice of making comfortable camps, and it is probable that the wonderful way they held out under their privations was largely due to this. While many in their predicament would have thrown away their blankets, they kept them, and on every cold and stormy night congratulated themselves that they had done so.

[Loss of boat] On Saturday, Aug. 1st, the first accident happened. Tracking on the Horse Shoe Rapids was extremely difficult and dangerous. Shortly after dinner a carry was made, taking three and a half hours to track out a path up and along a terrace about fifty feet high. Shortly after this the boat used by Cary and Smith capsized, emptying its load into the river. The party were "tracking" at the time, Cole being nearly the length of the tow line ahead, tugging on it, while Cary was doing his best to keep the boat off the rocks. At the margin of the swift unbroken current there were strong eddies, and in hauling the boat around a bend her bow was pushed into one, her slight keel momentarily preventing her from heading up stream again, and the rush of the water bore her under. At the same time Cary was carried from his footing and just managed to grasp the line as he came up and escape being borne down the stream. When things were collected and an inventory taken of the loss, it was found to include about one-fourth of the provisions, the barometer and chronometer rendered useless and practically lost, measuring chain, cooking utensils, rifles with much of the ammunition, axe and small stores, such as salt, sugar, coffee, etc. The loss was a severe one, and arose from failure to fasten the stores into the boats before starting, as had been ordered. The time given the party for the trip was so short, the distance so uncertain, and the things they desired to have an opportunity to do on the return that would require comparative leisure were so many, that they begrudged the few minutes necessary to properly lash the loads into the boats, each time they broke camp; and delay and disaster were the results. As the day was nearly spent, camp was made but about a mile from the last, and time used in repairing damages. A very ingenious baker for bread was contrived by Cole from an empty flour tin, a new paddle made to replace the one lost, and a redistribution of the baggage remaining effected.

In the following five days sixty-six miles were made with a few short carries, some rowing and a good deal of hard tracking. Having passed the Mininipi river and rapids, the latter being the worst on the river, the bank furnishing almost no foothold for tracking the Mauni rapids were reached and finally at 5 P. M., Aug. 6th, the party emerged into Lake Waminikapo. As Cary's journal puts it, here the party "first indulged in hilarity. " The hardest part of the work was over and had been done in much less time than had been expected. According to all accounts the falls should be found only thirty miles beyond the head of the lake, which is forty miles long and good rowing water, and about three weeks time yet remained before they were due at Rigolette. Added to this a perfect summer afternoon, comparatively smooth water, running around the base of a magnificent cliff and opening out through a gorge with precipitous sides, showing a beautiful vista of lake and mountain, with the knowledge of rapids behind and the object of the trip but a short way ahead and easy travelling most of that way, and we may readily understand why these tired and travel worn voyagers felt hilarious. Cary says of the scene: "As we gradually worked out of the swift water the terraces of sand and stones were seen to give way and the ridges beyond to approach one another and to erect themselves, until at the lake's mouth we entered a grand portal between cliffs on either hand towering for hundreds of feet straight into the air. And looking beyond and between the reaches of the lake was seen a ribbon of water lying between steep sided ridges, over the face of which, as we pulled along, mountain streams came pouring. "

One day was used in making the length of the lake, and at the camp at its head Young and Smith turned back. A very badly swelled hand and arm caused by jamming his thumb had prevented Young from getting any sleep and threatened speedily to become worse. This in connection with the loss of provisions in the upset made it expedient to send the two men back. The returning party was given the best boat, the best of the outfit and provisions for six days, in which time they could easily reach the mouth of the river. Meantime Cary and Cole pushed on into what was to prove the most eventful part of their journey.

The lake is simply the river valley with the terraces cleaned out, and was probably made when the river was much higher, at a time not far removed from the glacial period. The head of the lake is full of sand bars and shoals, much resembling the mouth of the river as it opens out into Goose bay. On both sides of the lake mountains rise steeply for one thousand or twelve hundred feet. Its average width is from two to three miles and it has three long bends or curves. Only one deep valley breaks the precipitous sides, but many streams flow in over the ridge, making beautiful waterfalls.

The river as it enters the lake is about half a mile wide, but soon increases to a mile. Twenty miles were made by the advance the day the parties separated, and at night, almost at the place where the falls were reported, nothing but smooth water could be seen for a long stretch ahead. Sunday, the 9th, twenty-five miles were made the good rowing continuing, by burnt lands, and banks over which many cascades tumbled. Monday, the last day's advance in the boats was made, the water becoming too swift to be stemmed, This day Cary got the second ducking of the trip—a very good record in view of the roughness of the work and the smallness of the boats. During this and the day previous an otter, a crow and a robin were seen. As a rule the river was almost entirely deserted by animal life.

[Mount Hyde] The next day the boat and the provisions, excepting a six days supply carried in the packs, were carefully cached, and at 10:45 camp was left and the memorable tramp begun. Each man carried about twenty-five pounds. The stream was followed a short distance, then the abrupt ascent to the plateau climbed, old river beaches being found all the way up. Ascending a birch knoll, the river was in view for quite a long distance and a large branch seen making in from the west. To the north the highest mountain, in fact the only peak in the vicinity, was seen towering up above the level plateau. Towards this peak, christened Mt. Hyde, the party tramped, and arriving at the top saw the country around spread out like a map. Way off towards the northwest a large lake was seen from which Grand River probably flows, and nearer was a chain of small, shallow and rocky ponds. The country is rocky, covered with deep moss and fairly well wooded, with little underbrush. The wood is all spruce save in the river valleys where considerable birch is mixed in. The black flies were present in clouds, even in the strong wind blowing at the top of Mt. Hyde, and made halt for rest or any stop whatever intolerable. Leaving the mountain, after taking bearings of all the points to be seen, the party struck for the river and camped on the bank between the two branches coming in from the westward, several miles apart. The following day, with faces much swollen from fly bites of the day before, the line of march was along the banks till 2 P. M. when the upper fork was reached.

The course of the river is southeast. This branch course is from the northwest. The main stream turns off sharply to the northeast and after a few miles passes into a deep canon, christened "Bowdoin Canon, " between precipitous walls of archeac rock from six hundred to eight hundred feet high. This canon was afterward found to be about twenty-five miles long and winding in its course. In but few places is the slope such as to permit a descent to the river bank proper, and the canon is so narrow, and the walls of such perpendicular character, as to make the river invisible from a short distance. It might truly be said that the discovery of this canon, infinitely grander on account of its age than any other known to geology, and surpassed by few in size, is the most important result of the expedition. Several photographs of it were made, which were not injured by the exposure to wet and rough usage that the camera had to receive during the return journey, and alone convey an adequate idea of this most wonderful of nature's wonders.

At night the first camp away from the river was made, on the plateau. The two men felt that the next day must be their last of advance, so weakened were they by the terrible tramping over deep moss and the persistent bleeding by black flies. The stock of provisions, too, was running low, and with their diminishing strength was a warning to turn back that could not be neglected. A half dozen grouse, three Canada and three rough, had been added to their supplies, but even with full meals they could not long stand the double drain upon their strength.

In the morning a high hill was seen, for which they started, drawing slightly away from the river. Soon a roar from the direction of the river was noticed, which differed from the ordinary roar of the rapids. Altering their course it was found the roar "kept away, " indicating an unusually heavy sound. Pushing forward, thinking it must be the desired falls, they soon came out upon the river bank, with the water at their level. This proved the falls to be below them, and looking down they could be seen "smoking" about a mile distant. A distinct pounding had also been felt for some time previous, which further assured them that the falls were at hand. The roar that had attracted their attention was of the river running at the plateau level. At the point they came out upon it, it was nearly two hundred yards wide, a heavy boiling rapid. Walking down the great blocks of rock which form the shore, the river appeared to narrow and at 11.45 A. M., the Grand Falls were first seen.

[The marked Bowdoin Spruce] After making pictures of the Falls a feeling of reaction manifested itself in Cary's physical condition, and he remarked, "I do not wish to go farther, I need sleep. " Cole, as assistant, had avoided the wear and anxiety of leadership. His athletic work at Bowdoin, in throwing the shot and hammer and running on the Topsham track, had given him stored energy of arm and leg. This reserve strength prompted him to press forward and see more of a region new to human eyes. Leaving his hatchet with Cary, now rolled up in his blanket, with the hope and expectation that on waking he would use the same in preparing fuel and cooking supper, Cole pressed forward into the strange and unknown country three or four miles, and then, for a final view of the location, climbed the highest tree he could find and from its top surveyed the waste of land and river. He stood thus exalted near the center of the vast peninsula of Labrador. Four hundred and fifty miles to the east lay the wide expanse of Hamilton Inlet. Four hundred and fifty miles to the north lay Cape Chudleigh, towards which he could imagine the Julia A. Decker, vainly as it proved, pointing her figure head through fog and ice. Only six hundred miles due south the granite chapel of Bowdoin College points heavenward both its uplifted hands. Four hundred and fifty miles to the west rolled the waves of that great inland ocean, Hudson's Bay, into whose depths, Henry Hudson, after his penetrations to northern waters above Spitzbergen, after his pushing along the eastern coast of Greenland, after his magnificent and successful exploration of the American coast from Maine to Virginia, penetrating Delaware bay and river and sailing up that river crowned by the Palisades and the hights of the Catskills, honored with his name and whose waters bear the largest portion of the commercial wealth of our own country; still fascinated by the vision of a northwest passage that intrepid explorer penetrated into the waters of the unknown sea whose waves unseen dash along the coasts of Labrador from its westward to its northern shores and Cape Chudleigh. All these explorations he accomplished in a sailing vessel about the size of the Julia A. Decker, the ship "Discoverie" of seventy tons. He had wintered at the southern extremity of Hudson's Bay surrounded by a mutinous crew. In the hardships and suffering of the next season, after he had divided his last bread with his men, in the summer of 1611, while near the western coast of Labrador, half way back to the Straits, by an ungrateful crew he was thrust into a sail boat with his son John and five sailors sick and blind with scurvy, and was left to perish in the great waste of waters, which, bearing his name, is "his tomb and his monument. " Cole, with his mind and imagination filled with these facts, involuntarily took his knife and carved his name and the expedition on the upper part of the tree which formed his outlook. It might be his monument as the Inland Sea was that of Hudson. Then to have the tree marked and observable to other eyes, in case other eyes should see that country, he commenced to cut the branches from near the top of the tall spruce. He regretted much the leaving of the hatchet with Cary as he was obliged to do the work with his knife. It was a slow and laborious job. His imagination, as it roamed over the wide land, and his interest in his present efforts, had consumed time faster than he knew, and the slanting rays of the western sun started him with thoughts of Cary and supper. It was dark when he reached Cary and he was still asleep. The hatchet was idle, and he wished more than ever that his efforts on the branches of the marked Bowdoin Spruce had been rendered less laborious and more expeditious by the aid of this, to be hereafter his constant companion and source of safety along with another and more diminutive friend, a pocket pistol.

[Grand Falls] The falls proper are three hundred and sixteen feet high, and just above the river narrows from two hundred and fifty to fifty yards, the water shooting over a somewhat gradual downward course and then plunging straight down with terrific force the distance mentioned, and with an immense volume. The river is much higher at times and the fall must be even grander, for while the party was there the ground quaked with the shock of the descending stream, and the river was nearly at its lowest point. At the bottom is a large pool made by the change of direction of the river from south at and above the falls to nearly east below. The canon begins at the pool and extends as has been described, with many turns and windings, for twenty-five miles through archaic rock. Above the falls in the wide rapids, the bed was of the same rock, which seems to underlie the whole plateau. In 1839, the falls were first seen by a white man, John McLean, an officer of the Hudson Day Co., while on an exploring expedition in that "great and terrible wilderness" known as Labrador. His description is very general, but he was greatly impressed with the stupendous height of the falls, and terms it one of the grandest spectacles of the world. Twenty years later, one Kennedy, also an employe of the Hudson Bay Co., persuaded an Iroquois Indian, who did not share the superstitious dread of them common among the Labrador Indians, to guide him to the thundering fall and misty chasm. He left no account of his visit, however, and in fact, though one other man reached them, and Mr. Holmes, an Englishman, made the attempt and failed, no full account of the falls has been given to the world, until Cary and Cole made their report. Above the falls as far as could be seen, all was white water, indicating a fall of about one hundred foot per mile. In the course of twenty-five or thirty miles there is a descent of twelve hundred feet, nearly equal to the altitude of the "Height of Land, " as the interior plateau of Labrador is called, which has probably been previously overestimated. The next forenoon was spent in surveying and making what measurements could be made in the absence of the instruments lost in the upset. At noon, after having spent just twenty-four hours at Grand Falls, the party turned back. The very fact of having succeeded, made distance shorter and fatigue more easily borne, so they travelled along at a rattling pace, surveying at times and little thinking of the disaster that had befallen them. Camp was made on the river bank, beneath one of the terraces which lined both sides.

Saturday Aug. 15th, the march back to the boat cache was resumed. Towards night, as they approached the place, smoke was seen rising from the ground, and fearing evil, the men broke into a run during the last two miles. As Cary's journal puts it: "We arrived at our camp to find boat and stores burnt and the fire still smoking and spreading. Cole arrives first, and as I come thrashing through the bushes he sits on a rock munching some burnt flour. He announces with an unsteady voice: 'Well, she's gone. ' We say not much, nothing that indicates poor courage, but go about to find what we can in the wreck, and pack up for a tramp down river. In an hour we have picked out everything useful, including my money, nails, thread and damaged provisions, and are on the way down river hoping to pass the rapids before dark, starting at 5. "

Their position was certainly disheartening. They were one hundred and fifty miles from their nearest cache, and nearly three hundred from the nearest settlement, already greatly used up, needing rest and plenty of food; in a country that forbade any extended tramping inland to cut off corners, on a river in most places either too rough for a raft or with too sluggish a current to make rafting pay; and above all, left with a stock of food comprising one quart of good rice, brought back with them, three quarts of mixed meal, burnt flour and burnt rice, a little tea, one can of badly dried tongue, and one can of baked beans that were really improved by the fire. Add to this some three dozen matches and twenty-five cartridges, blankets and what things they had on the tramp to the falls, and the list of their outfit, with which to cover the three hundred miles, is complete. There was no time to be wasted, and that same night six miles were made before camping. The next day the battle for life began. It was decided that any game or other supplies found on the way should be used liberally, while those with which they started were husbanded. This day several trout were caught, line and hooks being part of each man's outfit, and two square meals enjoyed, which proved the last for a week. A raft was made that would not float the men and baggage, and being somewhat discouraged on the subject of rafting by the failure, another was not then attempted, and the men continued tramping. Following the river, they found its general course between the rapids and Lake Wanimikapo, S.S. E. During part of that day and all the next, they followed in the track of a large panther, but did not get in sight of him. Acting on the principle that they should save their strength as much as possible, camps were gone into fairly early and were well made; and this night, in spite of the desperate straits they were in, both men enjoyed a most delightful sleep.

[Squirrel and Cranberries] After this some time every morning was usually occupied in mending shoes. All sorts of devices were resorted to to get the last bit of wear out of them, even to shifting from right to left, but finally Cole had to make a pair of the nondescripts from the leather lining of his pack, which lasted him to the vessel. Cranberries were found during the day and at intervals during the tramp, and were always drawn upon for a meal. About two quarts were added to the stock of provision, and many a supper was made off a red squirrel and a pint of stewed cranberries.

Wednesday, the 19th, another raft was made, which took the party into the lake. This was more comfortable than tracking, yet they were in the water for several hours while on the raft, which was made by lashing two cross-pieces about four feet long on the ends of five or six logs laid beside each other and from twenty to thirty feet long, all fastened with roots, and having a small pile of brush to keep the baggage dry. The still water of the lake made the raft useless, even in a fresh, fair breeze, and so this one was abandoned two miles down, and the weary tramping again resumed. Fortunately the water was so low that advantage could be taken of the closely overgrown shore by walking on the lake bed, and far better progress was made owing to the firmer footing. Three days were used in getting down the lake, during which time but one fish, a pickerel, was caught, where they had expected to find an abundance.

At the foot of the lake, tracks were seen, which it was thought might be those of hunters. It was learned later that they were more probably tracks of Bryant's and Kenaston's party, who were following them up and probably had been passed on the opposite side of the lake, unnoticed in the heavy rain of the preceeding day. Some bits of meat that had been thrown away were picked up and helped to fill the gap, now becoming quite long, between square meals. Supper on this day is noted in Cary's journal because they "feasted on three squirrels. " Having gotten out of the lake into rapid water, trout was once more caught, and as on the following day, Sunday, the 23d, a bear's heart, liver, etc., was found, and later several fish caught. The starvation period was over.

In the afternoon another raft was built and the next day carried them five miles down to the last cache. Though so terribly used up that the odd jobs connected with making and breaking camp dragged fearfully, and each day's advance had to be made by pure force of will, the men felt that the worst was over and their final getting out of the woods was a matter of time merely. At this cache, also, a note from Young and Smith was found announcing their passage to that point all right and in less time than expected, so they had drawn no supplies from the stock there.

Tuesday, the 25th. —The day, by the way, that the Julia Decker and party arrived at Rigolette according to plans, expecting to find the whole Grand River party, and instead found only Young and Smith, who had been waiting there about a week. Rafting was continued in a heavy rain down to the Mininipi Rapids over which the raft was nearly carried against the will of the occupants. At the foot of these rapids a thirty mile tramp was begun, the raft that had carried them so well for forty-five miles being abandoned, which took them past the Horse Shoe and Gull Island Rapids and occupied most of the two following days. The tracking was fair, and as starvation was over pretty good time was made.

Thursday, the 27th. —A raft was made early in the morning that took them by the Porcupine Rapids and landed them safely, though well soaked, at the head of the first falls. Camp was made that night at the first cache below the falls, forty miles having been covered during the day.

[The last pistol shot] Friday, they fully expected to reach Joe Michelin's house and get the relief that was sadly needed, but as the necessity for keeping up became less imperative, their weakness began to tell on them more. Cary's shoes became so bad that going barefoot was preferable, except over the sharpest rocks, and Cole's feet had become so sore that as a last resort his coat sleeves were cut off and served as a cross between stockings and boots. They were doomed to disappointment, however, and compelled to camp at nightfall with four or five miles bad travelling and the wide river between them and the house. Fires were made in hopes of attracting the trapper's attention and inducing him to cross the river in his boat, but as they learned the next day, though they were seen, the dark rainy night prevented his going over to find out what they meant. The last shot cartridge was used that night on a partridge, and the red squirrels went unmolested thereafter. This last shot deserves more than a passing notice. In one sense these shot cartridges for Cole's pistol were their salvation. Just before the expedition started from Rockland it was remarked in conversation that the boat crew under DeLong, in the ill-fated expedition of the "Jeanette", met their death by starvation in the delta of the Lena, with the exception of two, Naros and Nindermann, simply because their hunter, Naros, had only a rifle with ball cartridges, the shot guns having been left on board the "Jeanette; " that on the delta there was quite an abundance of small birds which it was almost impossible to kill by a bullet and even when killed by a lucky shot, little was left of the bird. Cole was impressed by these facts and upon inquiring ascertained that the pistol shot cartridges ordered by the expedition had been overlooked. He energetically set about supplying the lack, and after persistent search, almost at the last hour, succeeded in finding a small stock in the city, which he bought out. To the remnant of this stock which escaped the fire at Burnt Cache camp, as has been said, is the escape of Cary and Cole from starvation largely due.

The value of these cartridges had day by day, on the weary return from Grand Falls, become more and more apparent to the owner. At the discharge of the last one, the partridge fell not to the ground, but flew to another and remote cluster of spruces. To this thicket Cole hastened and stood watching to discover his bird. Cary came up and after waiting a little while, said, "It is no use to delay longer, time is too precious. " The value of this last cartridge forced Cole to linger. He was reluctant to admit it was wasted. In a few minutes he heard something fall to the ground, he knew not what it was, but with eager steps pressed towards the place, and when near it a slight flutter and rustling of wings led him to discover the partridge, uninjured except that one leg was broken; that by faintness or inability to hold its perch with one foot it had fallen to the ground. The darkness and rain of that night then closing around them were rendered less dark and disagreeable by the assurance that kind Providence showed its hand when the help of an unseen power was needed to deliver them from the perils of the unknown river. It rained hard all the next forenoon, and as the river was rough, the men stayed in camp, hoping Joe would come across, until noon, when a start was made for the house. A crazy raft took them across the river, the waves at times nearly washing over them, and landing on the other side, they started on the last tramp of the trip, which the rain and thick underbrush, together with their weakened condition, made the worst of the trip. About 3 P. M., they struck a path, and in a few minutes were once more under a roof and their perilous journey was practically done.

Seventeen days had been used in making the three hundred miles, all but about seventy-five of which were covered afoot. When they came in, besides the blankets, cooking tins and instruments, nothing remained of the outfit with which they started on the return except three matches and one ball cartridge for the revolver, which, in Cole's hands, had proved their main stay from absolute starvation. The following day, Sunday, after having had a night's rest in dry clothes and two civilized meals, Joe took them to Northwest River, where Mr. McLaren, the factor of the Hudson Bay Company's posts showed them every kindness till a boat was procured to take them to Rigolette. A storm and rain, catching them on a lee shore and giving the already exhausted men one more tussle with fortune to get their small vessel into a position of safety, made a fitting end to their experiences.

[On board the Julia A. Decker] Tuesday at 4 P. M., they reached the schooner and their journey was done. Amid the banging of guns and rifles, yells of delight and echoes of B-O-W-D-O-I-N flying over the hills, they clambered over the rail from the boat that had been sent to meet them and nearly had their arms wrung off in congratulations upon their success, about which the very first questions had been asked as soon as they came within hearing. They were nearly deafened with exclamations that their appearance called out, and by the questions that were showered on them. At last some order was restored, and after pictures had been made of them just as they came aboard, dressed in sealskin tassock, sealskin and deerskin boots and moccasins, with which they had provided themselves at Northwest River, ragged remnants of trousers and shirts, and the barest apologies for hats, they were given an opportunity to make themselves comfortable and eat supper, and then the professor took them into the cabin to give an account of themselves. It was many days before their haggard appearance, with sunken eyes and dark rings beneath them, and their extreme weakness disappeared.

The return trip of Young and Smith from Lake Waminikapo, who reached Rigolette Aug. 18th, was made in five days to Northwest River, and after resting two days, in two more to Rigolette. Their trip was comparatively uneventful. At the foot of Gull Island Lake they met Bryant and Kenaston, who with their party of Indians were proceeding very leisurely and apparently doing very little work themselves. At their rate of progress it seemed to our party very doubtful if they ever reached the falls. They had picked up, in the pool at the foot of the first falls, one of the cans of flour lost in the upset, some fifty or sixty miles up the river, with its contents all right, and strange to say not a dent in it, and returned it to Smith and Young when they met them. That night, with the assistance of the officers and passengers of the mail steamer, which lay alongside of us, a jollification was held. Our return race to Battle Harbor, the last concert of the Glee Club in Labrador waters, the exciting race over the gulf with the little Halifax trader, the tussle with the elements getting into Canso, the sensation of a return to civilization and hearty reception at Halifax, and greeting at Rockland, must remain for another letter.

* * * * *

ON BOARD THE JULIA A. DECKER,
ROCKLAND HARBOR, ME.,
September 23, 1891.

The staunch little schooner has once more picked a safe path through
the dangers of fog, rocks and passing vessels, and her party are
safely landed at the home port, before quite two weeks of the college
term and two weeks of making up had piled up against its members.

The crew that weighed anchor at Rigolette on the morning of
September 2nd, when the wind came and the tide had turned, was a
happy one, for from Professor to "cookee" we all felt that we were
truly homeward bound, and that we had accomplished our
undertaking without any cause for lasting regret. The mail steamer,
whose passengers had joined in the jollification of the night
preceding, being independent of the wind, had started ahead of us.
Another race was on with the "Curlew, " this time a merely friendly
contest, without the former anxiety as to some other party's getting
the lead of ours in the trip up the Grand River. But the result was not
different this time. A fine breeze kept us going all day and the
following night. But the next day the fog came. It was no different
from the cold, damp, land-mark obscuring mist of the Maine coast in
its facility in hiding from view everything we most wanted to see in
order to safely find the harbor that we knew must be near at hand,
though we could not tell just where. A headland, looming up to
twice its real height in the fog about it, was rounded, and the lead
followed in the hope that it would take us to the desired haven. Soon
a fishing boat hailed, and a voice, quickly followed by a man,
emerged from the fog and shouted that if we went farther on that
course we would be among the shoals. We were told we had passed
the mouth of the harbor, and so turning back, tried to follow our
guide, but he soon disappeared. Just at this moment when it seemed
impossible for us to find any opening, the fog lifted and we saw a
schooner's sail over one of the small islets that lay about us. Taking
our cue from that we poked into the next narrow channel we came
to, and getting some sailing directions from a passing boat, and from
the signal man stationed on a bluff to give assistance to strangers, we
glided into an almost circular basin, hardly large enough for the
vessel to swing in, set among steep rising sides, into which many
ring bolts were seen to be fastened, and perfectly sheltered from
every wind. The use for the ring bolts we found later. The fog kept
rolling over, and the little fishing vessels kept shooting in, till it

seemed the harbor would not hold another. As all sail had to be hauled down before the vessels came in sight of the interior, the vessels seemed literally to scoot into the basin. A few of the vessels were anchored and kept from swinging by lines to the bolts, and the rest of the fleet made fast to them. In all the number of vessels crowded into the space where we hardly thought we could lie was about twenty. How they would ever get out seemed a puzzle, but the next morning it was accomplished, with a light fair wind, by all at once without accident or delay. Had the wind been ahead, the ring bolts would have aided in warping to a weatherly position.

During the evening the mail steamer caught us, and after putting a little freight ashore, left us behind again. Here were some strange epitaphs painted on the wooden slabs, also people ready to exchange or sell at a far higher rate than we had hitherto paid, anything they possessed for the cash which was all we had left to bargain with, the available old clothes having been already disposed of.

It was hard to disabuse the minds of the people at Square Island Harbor of the idea that we had come to seek gold or other valuable mines, the reason being that several years before a party from the States had spent considerable time prospecting in that vicinity and partly opened one or two worthless mica quarries.

[A Bold Skipper] It was a glorious sight to see the fleet get under way the next morning. Many a close shave and more bumps but no serious collisions were caused by the twenty or more vessels crowding out together through the narrow opening, each eager to get the first puff from the fair breeze outside the lee of the cliffs. The whole fleet was bound up the coast, but before many of the schooners had drifted far enough out to catch the breeze it had failed, and only after an hour or more of annoying experience with puffs from every quarter, did the strong sea breeze set in. Sheets were trimmed flat aft, and all settled down to beating up the coast. The Julia soon left the mass of the fleet and before reaching Battle Harbor, where a long desired mail was awaiting, had nearly overtaken the lucky ones who had drifted far enough off shore to make a leading wind of the afternoon breeze. During the calm a school of whales disported themselves in the midst of the fleet, chasing one another, blowing and churning the water to foam about us, apparently as though it was rare fun.

Late in the afternoon we approached the entrance to Battle Harbor, but with the wind blowing directly out of the narrow, rocky and winding entrance we wondered how we should get in. Our captain was equal to the problem, however, and undeterred by the crowded state of the harbor, within whose narrow limits were two large steamers, one or two barks and several fishermen, performed a feat of seamanship the equal of which, we were told, preserved in the traditions of the port, and only half believed, as having been done once, thirty years before.

Getting about ten knots way on the vessel, and heading her straight for the steamer nearest the mouth, we just brushed by the rocks of the entrance, sheered a bit and shot past the steamer before her astonished officers could utter a word of warning, and were traveling up the harbor at a steamboat pace, the sails meanwhile rattling down, and some of us on board wondering if we should not keep right on out the other entrance to the harbor, while boats scurried out of our way, two men in one fishing boat looking reproachfully at us as we missed them by about two feet just after our fellow on lookout had reported "nothing but a schooner in the way, sir; " and people rushed to their doors and to the decks to see what was exciting such a commotion, just as the anchor was let go with a roar and we quietly swung to and ran our mooring line, as though we had done that thing all our lives.

Here about one hundred letters were brought aboard amid much rejoicing, for many had not heard from home at all during the trip.

By the time we were ready to make what we hoped would prove the last departure from a Labrador harbor, the next morning, the wind, which had changed in the night and was blowing in exactly the opposite direction, had become so strong that the little steam launch of Bayne & Co., which had been tendered us to tow us out of the harbor, was not powerful enough to pull the schooner against it. The other entrance, for like all the rest this Labrador harbor was merely a "tickle" and had its two entrances, was narrow, shoal, and had such short turns that it seemed impossible to run so large a vessel as the Julia through it. However, our impatience would not brook the uncertain delay of waiting for the wind to change, so taking on board the best pilot that town of pilots could afford, we made the attempt. Three times we held our breaths, almost, as we anxiously watched the great green spots in the water, indicating sunken rocks, glide under our counter or along our side, while the steady voice of the weatherbeaten old man at the fore rigging sounded "port, " then in quick, sharp, seemingly anxious tones, "now starboard—hard! " and again "port—lively now, " and the graceful vessel turned to the right or left, just grazing the rock or ledge, as though she too could see just how near to them it was safe to go and yet pass through without a scrape. It was a decided relief to all, and the silence on board, that had been broken only by the rush of wind and water, the pilot's voice and the creaking of the wheel as it was whirled around by the skillful hands of the captain, suddenly ceased, when the pilot left his place and walked slowly aft, praising the admirable way in which the vessel behaved at the critical points, and apparently unconscious that in the eyes of twenty college boys he had performed an almost impossible feat.

After a hard pull to windward for two of us, to set the pilot ashore, and a wet and rough time getting aboard again, and after our laugh at the expense of the mate, who had cast off our shore warp, as we started out of the harbor, and then had been unable to catch the schooner, which was equally unable to wait for him in the narrow passage, and who had, therefore, to row all the way after us at the top of his speed, and only caught us when we lay to to send off the pilot; we made everything snug and started down the straits, hoping to reach Canso without further delay.

[Last harbor in Labrador] That was not our fortune, however, for soon the wind hauled ahead, and with a strong current against us it was impossible to make any progress, so after jumping in a most lively manner all day, in the chops of Belle Isle, we made a harbor for the night at Chateau Bay, in almost the same spot where we had waited two dreary days two months before. The next day we worked along the coast, but at night again put in to what proved our last, as well as our first harbor on the Labrador—Red Bay. Here we found a mail steamer and were allowed irregularly to open the bag to Battle Harbor and take out that which belonged to us, much to our delight, of course, for it gave us news comparatively fresh, that is, not over a month old, from home.

Here, also, we laid in a supply of the only fruit that Labrador produces, called "bake apple. " It is a berry of a beautiful waxen color when ripe, otherwise looking much like a large raspberry, and having a most peculiar flavor, which we learned to like, and grew very fond of, when the berries were served, stewed with sugar. We had been deprived of fresh fruit so long that we should probably have learned to like anything, however odd its flavor, that had its general characteristics.

Here, too, we again fell in with our little Halifax trader, which gave us so hot a race to Halifax in the coming week, both vessels arriving at Halifax within an hour of each other, after starting at the same time from Red Bay and keeping within sight nearly all the time. At length the wind came to the south, and we started, laying our course west, along the Labrador shore, so as to get a windward position and be able to "fetch" Canso when the wind came around to the west, as it is certain to do at that season of the year, compelling us to "tack ship" and stand right out against the stormy Gulf of St. Lawrence. These southwesterly winds had been our dread, for they blow so strongly and in September make the Gulf so rough that getting to windward against them is impossible. Hence our satisfaction can be imagined as we sped along the Labrador coast that day, the wind becoming a trifle easterly, so as to allow us to "start our sheets" and at the same time steadily increase our offing, getting such a weatherly position for Canso that the moment the expected change of direction began we promptly "tacked ship" and at the worst had a leading wind across.

For three days we hobnobbed with the little "Minnie Mac" across the Gulf. The first thing we did in the morning was to hunt her up with the glasses from aloft, if not in sight from the deck, and the last thing in order at night were speculations as to where we should next see her. The difference in the build of the two vessels, the one being shoal and centerboard, the other deep and heavily laden, made the race a zigzag. When the wind favored a little and the sheets could be "eased" then the shoal model would push ahead, but when the wind came more nearly ahead, and we had to plunge squarely into a head sea, then the deeper draught and heavier lading told to advantage.

During this time we were not idle on board. The Grand River men were beginning to feel vigorous again, and their notes and data had to be worked up. The collections, too, though largely packed away securely for the rough voyage, yet gave plenty of occupation to those not otherwise employed, while the few really industriously inclined used their superfluous energy in seeing to it that the lazy were given no opportunity to enjoy their idleness.

The morning of the fourth day the coasts of Cape Breton were in sight, but the wind came straight out of the Gut of Canso in half a gale, and then our rival, owing to her greater weight, forged ahead, and it seemed that we were to be beaten. However, much to our amusement, when we got a few miles off the mouth of the Gut, we found a calm, into which the "Minnie Mac" had run and where she stayed till we came up. With us also came a breeze, and we forged ahead of her into the anchorage at Port Hawksbury just as we had said we would do when we left Red Bay. Here we spent the rest of the day, laying in a stock of much needed fresh provisions, and sending nine of our college base-ballists, at the invitation of the Port Hawkesbury nine, to give them some points on the game. About the fifth inning the game closed on account of darkness, with score in Bowdoin's favor something about 30-0.

A short run brought us into Little Canso, where we had to turn to the west to go along the Nova Scotia coast to Halifax, but fog shut down so we spent a day inspecting the plant of the Mackay-Bennett cable, which has its terminus at Hazel Hill, about two miles from Canso, finding some very agreeable acquaintances in the persons of Mr. Dickinson, the manager, and Mr. Upham, his first assistant electrical expert, who proved to be a Castine man and was deligted to meet some Yankees from his old cruising grounds, Penobscot Bay, and getting some interesting knowledge concerning ocean telegraphy. It seemed strange, to say the least, to be in communication, as we were, with a ship out in mid-Atlantic, repairing a cable, and to have an answer from Ireland to our message in less than a minute after it was sent.

[Solid shot at Halifax] With one stop on account of fog and threatening storm, we reached Halifax in two more days. The introduction to it, though, was not so pleasant, for as we were running up the harbor solid shot from one of the shore batteries came dropping around us and skipping by us, altogether too near for comfort. However, no damage was done beyond the injury threatened to Her Majesty's property in the proposition for a while considered to call away boarders, land and take the battery. We found later that it was merely target practice and nothing disrespectfully intended towards the flag flying from our peak, so were satisfied that we had not made any hostile response.

Once ashore the hospitable Haligonians began by inviting the Professor and others to a dinner at the Halifax Club. The next day we enjoyed an official reception, and accompanied by Premier Fielding and members of his Cabinet, Consul General Frye and other gentlemen, were taken on an excursion about the beautiful harbor in the steam yacht of one of our entertainers, given a dinner and right royally toasted at one of the public buildings, and were finally taken to the Yacht Club House for a final reception.

At Halifax some of our party fearing more delay in reaching Rockland, left us, so with diminished numbers but plenty of enthusiasm we made ready for the last stage of the voyage. After some rather amusing experiences with our assistant steward or "cookee, " who seemed to reason that because he had been so long deprived of the luxuries of modern civilization he should employ the first opportunity he had to enjoy them in making himself incapable of doing so, and who was brought aboard the morning we sailed only after a somewhat prolonged search, we "squared away" for Cape Sable. The fine fair wind ran us nearly down there, but just as we thought to escape the provoking calms that delayed us in this vicinity on the outward trip, we found the wind drawing ahead and failing. A day was spent in slowly working around the cape, drifting back much of the time, and then we struck one of the southerly fog winds that are too well known on the Maine coast. We were in waters on which our captain had been bred, and so we pushed on into the night, looking eagerly or listening intently as the darkness closed over us for some sign of approaching land. At length, just about eleven, when it seemed we could not stand the suspense of knowing that thousands of rocks were just ahead but not just where they were, and yet equally unwilling to stop then, when so near home, we heard the sound of the breakers, and standing cautiously in on finding the water very deep, soon made Mt. Desert rock light. It was a welcome sight, and from there an easy matter to shape our course for home. At day-break we could still see nothing, but towards noon, the wind being light and our progress slow, we passed the desolate house of refuge on the Wooden Ball Island, and soon the lifting fog showed us the mouth of Penobscot's beautiful bay, and shortly after we dropped our anchor in the long wished for Rockland harbor, and the cruise of the Julia Decker and her crew of Bowdoin boys was ended.

[The royal welcome] The account would be incomplete, though, were reference omitted to the royal welcome that awaited us at Rockland. Upon landing we found the church bells ringing, and the city's business for the moment stopped, while the city fathers as well as a goodly number of her sons and daughters greeted us at the wharf. In the evening there was another reception, and there the expedition as such appeared for the last time, and as the most fitting way in which we could express our gratitude at the interest shown in our work and safe return, as well as to contribute our share towards the evening's entertainment, the Bowdoin College Labrador Expedition Glee Club rendered, as its last selection, a popular college song, of which the burden was, as also the title, "The wild man of Borneo has just come to town. "

JONATHAN P. CILLEY, JR.

* * * * *

[Missionary in Labrador] Since the Bowdoin College Labrador Expedition much interest has been taken by charitable women in the missionaries who are laboring in that bleak country. As often as possible barrels of clothing and other useful articles have been sent to them. In return the missionaries have sent interesting letters describing their work and acknowledging the gifts. One of these, written to Mrs. James P. Baxter, of Portland, gives a description that will be of general interest:

HOPEDALE, LABRADOR,
October 3, 1893.

Dear Madam:

For your very kind letter and for the very useful articles for our people, accept my best and kindest thanks. We have already made some of the people glad with cloth, and we will but be so glad for them in the winter time.

Happily the codfishery has been much better this year than last, thus we can more confidently look forward to the coming winter time than we could last year; because our people were so poor and we finished the many kind gifts long before the spring came on, when they were able to earn their own bread.

We have had a very cold and dreary summer, the few warm days could easily be counted, and now the winter is at the door.

On last Christmas day we had a nice Christmas celebration with our school children in the chapel. For this purpose we had placed two nice Christmas trees and two illuminated transparents in the chapel. My dear husband translated some lovely Christmas songs into Eskimo, and I taught the children to sing them. Between the hymns they recited songs and texts from the Bible. Sometimes one by one and then again altogether. The children made it very nicely. The choir, which sang some nice pieces, helped to make the whole to sound better. Finally every child got a large biscuit and a cup of tea, which seemed to make greater impression than the whole celebration. The congregation were also invited and they were very much interested in it.

In the midst of February I accompanied my dear husband on his journey around to the settlers belonging to our congregation, which live scattered far away from here towards the South.

We left Hopedale one morning, having 30 degrees Cen. of cold, of course by "kamatik" (dog sledge). I was well wrapped up so that I did not freeze so very much, but the worst is always on such a trip that we cannot eat anything. Before we started I made some meat balls for the purpose to use them during the nine hours driving, but it was impossible to make use of them because they were like stones without fearing to loosen our teeth. Happily I had some biscuits and to become more strengthened I used a little chocolate. We were nearly three weeks away from home and in that time we were nearly every day on the kamatik. Never less than five hours at a time, but generally from seven to nine hours, and twice from eleven to twelve hours. It was indeed sometimes very exhausting especially one time when we came to very poor people where we had for two days nothing to eat and the next day we had to travel for about eleven hours having nothing but dry biscuits. I did not feel so very well that time.

Many of these settlers have only the opportunity once a year to hear the gospel of God preached to them, that is when the missionary is visiting them. Many are too far away from Hopedale to come and visit us, and some are too poor; or at least the dogs' food is too expensive. My dear husband made this journey last winter for the fifth time, that is only towards the south. To the north he has also been different times. In such a journey the Sacraments are spent, marriage performed, and meetings are kept as many as possible. The poor children who grow up without having any school are examined as to how much they have improved since the last year. We felt this year very much again the need of having a station among them. There are children among them from 16 to 17 years of age who cannot read at all. We have now asked our society in London and Berthelsdorf, if possible, to build a station for them that they may have their own minister and teacher. We hope it may be done, then we would not have to travel any longer only in cases of need. Every one who has to travel ruins his health if he has to do it for a long time. The settlers could then easily reach the Mission Station or the missionary could in one day get to the place where he is wanted.

[Hungry children] May I, dear madam, give you some instances? First about a family having ten children of ages ranging from two to eighteen years. We came to that place in the afternoon about 5 o'clock accompanied by four other persons belonging to their relationship who joined when we left their homes. As soon as we opened the door of the house we were in the dwelling room. At the first sight we saw that great poverty governed here, even the children looked consumed and clothed in rags. The house was so bad that the wind made its way through the many gaps. After I had wrapped myself in a large shawl and placed myself beside the big stove I was still freezing. Some windows were broken, the opening filled with rags. My dear husband asked why they had not nailed a board on the place instead of rags; they answered, "We have got none. " But my husband said "You could easily have made a nail of wood, " which they promised to do. We could only get a very little bread, because they had only one small piece. I gave the tea. My dear husband spent the Sacrament, communion and baptism in the evening in the hope we would be able to go further the next day, for we could not stay any longer here if we would not starve. We had a poor resting place. It was not possible to undress ourselves. The whole time we felt the snow on our faces and the wind through the many gaps. We froze very much although the fire was kept on during the night. Not very far from us Mr. and Mrs. Tacque were resting, and we heard how the one said to the other, "I hope Mr. and Mrs. Hansen can go further to-morrow, for we have nothing to eat. " That was indeed a very sad prospect, for we heard too well the snow storm was howling outside and there was no hope for us to go on. And so it was. The next day I gave from our provisions as much as I could, but we had not very much, and I could not give everything away because we might afterwards be caught out in a snowstorm, which often happens, where we then have to live in a snow house until the storm is over. I gave now coffee for 19 persons, bread we had none, for it always freezes so hard that it is useless. The poor woman collected all the bread she had and we took as little as possible. During the day time my dear husband kept different meetings, talked and prayed with them. For dinner I asked for a large pot and put it on the stove. I had happily taken some preserved soups and cooked now for all the people in the house, put all our meat balls and broken biscuits into the same pot, and gave now from this dish a plateful to every person in the house. I had also put some "Liebig" in my box, before I left my home, and was now able to make the best use of it. It was something touching to see the many hungry children, how they devoured their portion. Anything like that they have perhaps never tasted before, and would gladly have taken some more, but it was already gone. In the afternoon my dear husband kept school for the children, told nice stories and instructed them about different things, and the children would have gone on for a long time. The smell in the house was not so very pleasant, 19 persons in one room, beside this the men smoked their pipes nearly the whole time. The children were crying and would not obey their parents and the parents are so very weak in this way.

In the evening I gave once more what I possibly could spare, and for the next morning too. But we really did hunger.

The Lord heard our prayers that we were able to go on the next morning to the next place, but because of the deep snow we could only move on very slowly. First after 11 hour's travelling we came in the evening to our next station. We did hunger more in these three days than we have done in our whole lives. The next place was a nice clean house, where we restored ourselves again.

In one place we visited an Eskimo. When we entered the room, what did we see? A seal living in the midst of their room. The people had heard of our coming and thus put the monster in the room to thaw it up to feed our dogs with. The animal was soon taken away. The house was clean, but small. In this place we had to sleep on the floor, and we used our blankets to make a couch as well as we could. A sailcloth was used as a curtain, so that we had something like a separated place for us. Our two drivers were also in the same room, and they cared for music during the night, for they snored like a saw mill, and when they woke up they smoked their pipes and gave the air in the room such an odor, which I shall not try to describe. Nevertheless, for all that, we were happy together, and I did not repent one minute to have accompanied my dear good husband, in order to be a faithful partner to him. We remembered also it was not a pleasant, but a mission trip we made, where we may expect many things like that. What is that little we can do for our Lord and Saviour? It is like a drop of water in the bottomless sea of his love. If our journey has but been a blessing to some, and if here and there one corn of gospel's seed may grow up we are more than paid for.

[Easter] We had four nice places where the good people did all they could to make it comfortable for us. Everywhere they were very thankful for my coming, and expressed their gratitude in many ways. At Easter time we had more visitors than usual and they seemed to be more happy than else.

Will you kindly excuse this short description, dear madam; it would take me too long to describe the whole journey. I used some of your kind gifts for the people whom we visited, and I hope you will, dear madam, and the kind ladies who contributed to your large and rich sending accept our and the people's warmest and best thanks.

With kindest regards from my dear husband and me, I am, dear madam, believe me,

Your affectionately,
ANNIE HANSEN.